KOREAN REGIONAL FARM PRODUCT AND INCOME: 1910-1975

by

Albert Keidel, III

1 9 8 1

KOREA DEVELOPMENT INSTITUTE

Seoul, Korea

All Rights Reserved by
THE KOREA DEVELOPMENT INSTITUTE
P.O. Box 113, Cheong Ryang, Seoul, Korea

Distributed Outside Korea by
The University Press of Hawaii

ISBN 0-8248-0758-8

"TO THE MEMORY OF MARIE BERGER"

FOREWORD

The importance of agriculture for a nation's overall economic development has been appreciated for some time, but the interrelations between industry and agriculture are still imperfectly understood. The usual emphasis in studies of these interrelations is on transfers of resources from agriculture to industry, or on the provision by industry of modern inputs to raise rural productivity. The present study compiles and analyzes voluminous regional data from South Korea's rural development to show that raising rural productivity involves more than the mere application of modern inputs.

The summary of Korean experience reported here argues that there is an intangible symbiotic relationship between industrial activity and farm productivity that transcends traditional input variations. The results are important for development policy, for they imply that the degree of industrial decentralization may significantly influence the degree of rural development success. In economies with regional inequalities or isolated rural poverty, or indeed in economies where the major concern is only to raise farm output with minimum sacrifice, these data may provide insights for broad strategy alternatives.

For this reason, the Korea Development Institute is pleased to be able to present this original study of South Korea's regional farm development. At the same time, we feel the study should be of general interest and value to students of South Korean agricultural history, for it presents for the first time an account of provincial development patterns, providing a rich extension of work already done at the national aggregate level.

The author, Albert Keidel, conducted all of the research while in residence at KDI. Although we have encouraged and supported his research at different stages, the views expressed in this study do not necessarily reflect the official position of the Institute.

Mahn Je Kim
President
Korea Development Institute

ACKNOWLEDGEMENTS

The present study is the result of eighteen months of research in Seoul, Korea and was presented to Harvard University in April, 1978 as the author's Ph. D dissertation. A special debt is owed to Dwight H. Perkins, patient teacher and generous adviser for over seven years. He developed the author's initial interest in Korea and also provided the opportunity for its pursuit.

The author benefited immeasurably from the year and a half spent at the Korea Development Institute (KDI) in Seoul. Through a generous agreement between its President, Mahn Je Kim, and the Harvard Institute for International Development (HIID), the author was given access to materials, computer resources, and the expertise of KDI's excellent professional staff. Particular thanks are appropriate for advice and criticism received from Research Director Kwang Suk Kim and from Senior Fellows Sung Hwan Ban, Pal Yong Moon, and Byung Nak Song. Sa Hon Kim provided valuable information from his own research. Kyu Soo Kim and Yu Il Kim of KDI's Statistical and Computer Division taught and assisted the author with most of the programming used in the analysis.

Researchers and planners in the Korean government were extremely helpful. In particular, Jong Woo Nam and Shin Yoon of the Crop Statistics Division, Ministry of Agriculture and Fisheries, Yoon Taek Chung of the Highway Planning Division, Ministry of Construction, and Sung Ho Kim of the National Agricultural Economics Research Institute were generous and patient in response to the author's many requests and questions. In Tokyo, Shigeru Ishikawa of the Economic Research Center, Hitotsubashi University kindly gave the author a complete set of provincial data worksheets from his colonial period research. Finally, David Cole of HIID provided timely guidance and advice at several important stages in the research.

The author wishes to thank Susan Dunham of Cambridge, Massachusetts for her excellent job typing the final manuscript.

Warmest thanks, however, are reserved for Chul Joo Kim and his wonderful family in Jang Ui Dong, who for eighteen months accepted the author into their home as one of their own and through whom the author feels a bond with the Korean people and the subject of his research which can never be broken.

Albert Keidel

Cheongryang-ri, Seoul
July 1979

TABLE OF CONTENTS

APPENDICES

BIBLIOGRAPHY AND INDEX

LIST OF TABLES

xv

LIST OF MAPS

CHAPTER I

INTRODUCTION: THE INFLUENCE OF URBAN CENTERS ON RURAL DEVELOPMENT

The farming industry in a poor country has long been recognized as a vital source of potential for national growth, both because of its usually overwhelming presence in the economy and because of the promise it holds as a source of the manpower and physical resources needed by other smaller but much more rapidly expanding sectors. A great deal of attention has been paid to this interaction and to the contributions made by agriculture in the process of modern industrialization. There has been less research, on the other hand, concerned with the influence of nascent or burgeoning industrial centers on the outlying rural economy. The present study traces the history of agricultural development in separate provinces and regions of South Korea over the course of the twentieth century and in so doing uncovers a pattern of farm productivity response to contact with cities which transcends the usual relationship between yields and the improvement of inputs and measurable technique.[1]

The goals of the study are several. In the first place, it is important for its own sake to detail the regional pattern of Korea's rural

---

[1] An ideal methodology would have compared suburban and outlying rural farm areas over the entire period to capture more accurately the influence of urbanization on farm incomes. The techniques employed for the historical portions in this study, however, are based on large provincial units and are hence much less precise. The detailed 1970 county results reported here are quite precise in this sense, but are for only one year. The former method is used over time in a limited way by Yang-bu Chai, "A Study on the Regional Inequality of Farm Income in the Process of Urban-Industrialization in the Seoul Area," M. S. Thesis, Seoul National University, 1971.

growth. The national trends in agriculture's growth have been well studied using statistics for the country as a whole, but has this aggregate process been reflected evenly in the experience of individual provinces? In spite of a certain degree of homogeneity throughout rural Korea, each region has followed a unique path reflecting natural and man-made differences in overall endowment. Some areas have been in the vanguard of general nationwide trends while others have lagged behind, and a regional study allows a much more careful mapping of Korea's overall farm experience. For example, in many economies regional differences in average farm income form an important component of overall rural income distribution, and it is interesting to find that this element, although significant for South Korea, is not extremely so.

In the second place, a study of the farm history of individual provinces permits a much more detailed investigation of the forces responsible for agricultural growth than is the case when only national statistics are relied on. What has been the significance of fertilizers, pesticides, irrigation, new seeds and even weather for the expansion of rural output? Why do some regions stagnate while others surge ahead, pulling with them the overall farm economy? The regional record shows that both traditional inputs such as those listed above and exotic factors related to urban industrial growth play important and measurable roles in the overall process of rural development.

Finally, the wealth of material provided by regional sources makes us bold enough to ask if the experience of rural Korea during some 70 years of growth holds lessons for other farm sectors in economies starting to grow. The regions of South Korea are comparable in many dimensions, and yet some are on average richer than others. Are the discernable reasons for this divergence sufficiently general to warrant their application in other cultures and climates? There is little new information in the confirmation of the central importance of chemical fertilizer, irrigation and new seeds, but the clear significance of links to cities demonstrated by Korea's experience is an overarching consideration perhaps too often overlooked or underestimated in strategies for rural development.

The statistics used to deal with the above issues are in general the product of Japanese colonial administrative discipline, which collected and compiled data of all kinds for Korean provinces, counties and even districts and recorded them in a framework which remained essentially unchanged throughout both the colonial period before World War II and the post-liberation period after 1945. A major contribution of this study has been the hunting down and collection of these regional statistics, often obtainable only in old, obscure and even disintegrating publications, and their rendering in consistent units and in a time-series format. The backbone of the study's statistical framework is the set of reported output levels for individual crops by province from 1910 to 1975, and though there were only several dozen such crops in the Japanese colonial era, the number of individually recorded outputs was well over 70 by the end of the period covered. In addition, provincial time series on farm population, land holding, fertilizer consumption, farm machinery, irrigation and a host of other variables present enormous potential for analysis and research.[2] The work in the present study represents only a first pass at interpreting the lessons these data hold for students of Korean agriculture.

Although the provincial time-series data thus provide a wealth of historical information, regionally more detailed statistics for individual counties in the single year 1970 form a supplementary body of data which greatly enriches the conclusions drawn from provincial sources. Korea has 170 counties but only 9 provinces, and where the larger units do not always conform with natural or economic regions, the county division is fine enough to permit the confirmation and elaboration of otherwise much more tentative results. Some of the 1970 county data are administrative, but the bulk are from an agricultural census taken in the same year as part of the decennial World Census of Agriculture carried out under the auspices of the United Nations. It is in fact these more detailed data which provide the most dramatic exposure of the impact of urban centers on farm productivity.

What are the specific findings gleaned by this study from such a maze of information? The most clear-cut conclusions are historical and descriptive. At the beginning of the twentieth century the regions of Korea were in a state of rough equilibrium. Differing farm population densities compensated for different levels of soil productivity, and per-household output was fairly well balanced from one province to the next. Over the course of 35 years under Japanese control, however, one region in the Northwest disrupted this balance by

---

[2] See Appendixes A and B for a more complete discussion of these data and their quality.

rapidly increasing its average farm output, and another region in the Northeast became very poor by the same measure when it seriously failed to keep up with the pace of South Korea's overall rural growth. This pattern of regional inequality which emerged before World War II remained the overarching pattern after 1945, but between the two extremes important regional shifts in per-household product allowed provinces with previously underutilized land to dramatically eclipse areas traditionally more affluent.

A variety of inputs contributed to both the overall increase in South Korea's twentieth-century rural output and the pattern of increased regional inequality, but the most significant of these was chemical fertilizer, which is not surprising when one considers the centuries of intense cultivation of South Korea's limited land and the inability of traditional compost and manure treatments to fully compensate for the accumulated nutrient loss. Other important factors have been improved seeds, irrigation, vinyl greenhousing and pest and disease control. An important explanation for basic regional differences is land endowment, and the 1970 county data show that paddy (irrigated and not), dry fields and orchards contribute differently to farm output.

The greatest single factor explaining regional differences, however, is the influence of Seoul, Pusan, Taegu, Incheon, Taejeon and other major urban centers on the nearby agricultural belts. This conclusion is very clear from provincial data in both the Japanese and post-liberation periods, and detailed analysis of the isolation of rural counties in 1970 from 14 industrial centers reveals that every kilometer of distance from such centers significantly reduces value-output for a representative farm. In this respect, transportation, and in particular roads, have been extremely valuable for the transmission of urban stimulation. The cropping patterns and yields of one province underwent striking transformation and acceleration after the construction of a section of limited access highway linking it with Seoul.

The importance of isolation and the benefits of urban proximity, then, stand out as the most salient lesson to be learned from the regional time-series and county data gathered for this study. Such a finding is by no means original or new, although examples of its demonstration for developing countries have been few and less detailed than the work presented in this study for South Korea. The general concept of urban influence on rural areas has nevertheless been clearly stated by economists and persuasively proven by historical analysis of farm communities in the United States. The following passages will briefly present those generalizations and the results of empirical studies; such studies, better than any other words, serve as a proper introduction to the chapters and appendixes which follow.

The relevance of spatial relationships is familiar in economic literature, is usually dealt with in the context of "growth centers" (as opposed to "growth poles")[3] and relies heavily on location theory and the importance of transportation costs for the location and success of industries. The treatment of agriculture, however, seems to have been developed independently in the work of Theodore W. Schultz as he investigated poverty and the value of land in rural America; the empirical studies reviewed below were all carried out within his theoretical framework.

Schultz's framework is based on two sets of propositions. The first set appeared in 1950[4] and states that (a) before and at the beginning of industrialization, incomes in farm communities are more equal than they are after industrialization has proceeded; (b) eventual differences in farm incomes result from some communities becoming better off rather than others becoming poorer; and (c) changes in farm community incomes result from changes in the overall economy and not from initial differences in land, cultural values or human talent. Beginning with the statement that development brings about economy-wide disparities in income, Schultz concludes that these disparities result in parallel farm income disparities and at the same time create impediments to factor-price equalization, in particular impediments to migration.

Schultz's second set of propositions appeared in 1951 and was expanded in 1952.[5] These propositions state that (a) all economic

---

[3] See D. F. Darwent, "Growth Poles and Growth Centers in Regional Planning: A Review," *Environment and Planning,* Vol. 1, 1969, pp. 5-31. Reprinted in J. Friedman and W. Alonso (eds.), *Regional Policy: Readings in Theory and Applications,* Cambridge: M.I.T. Press, 1975, for a good review and a critique of many of the spatial and growth pole models of analysis.

[4] Theodore W. Schultz, "Reflections on Poverty within Agriculture," *Journal of Political Economy,* Vol. 58, No. 1, February 1950, pp. 1-15. Reprinted in American Economic Association, *Readings in the Economics of Agriculture,* Homewood: Richard D. Irwin, 1969, pp. 321-338.

[5] Theodore W. Schultz, "A Framework for Land Economics—The Long View," *Journal of Farm Economics,* Vol. 33, No. 2, May 1951, pp. 204-215; Theodore W. Schultz, *The Economic Organization of Agriculture,* New York: McGraw-Hill, 1953, pp. 146 ff.

development occurs within a spatial or locational matrix; (b) the economic development referred to is industrial and urban in nature; and (c) economic organizations work best at or near the centers of these locational matrixes and hence also work best for farms situated close to such centers. He illustrates these points with the fact that nowhere within the milksheds of urban-industrial areas are there to be found poor farm communities, regardless of whether the land is hilly and poor or flat and fertile. Capital markets work better for the farmer in urban areas, and the increased site value of the farmer's land, unrelated to its agricultural qualities, biases the capital market in his favor.

Schultz's propositions form a set of testable hypotheses about rural development and its relationship to economic modernization. Researchers have explored not only the correlation between urban-industrialization and farm incomes, but also the relative strengths of linkages via capital, labor and other markets. A particularly useful set of circumstances was provided by the Tennessee River Valley, which in the first half of the twentieth century went through a process of industrial growth much faster than that of the United States as a whole. Two of the studies outlined below tested Schultz's hypotheses using data from counties in the Tennessee Valley.

Ruttan[6] used United States census data to show that the positive relationship between median farm income and the level of urban-industrial development is almost as strong as the corresponding relationship for non-farm income. In other words, farm communities in the Tennessee Valley region located near centers of industrial activity had consistently and significantly higher median farm income than more isolated farm communities. Ruttan also attempted to trace these differences to the influence of markets for (a) capital, (b) labor, (c) farm product, and (d) intermediate inputs. His analysis, however, was weak for the latter two markets and concentrated on markets for labor and capital. Labor benefitted most from the availability of non-farm employment, while the data on capital inputs into agriculture showed a much higher level near the industrial centers. Ruttan's results are particularly interesting for questions concerned with cultural origins of poverty in agriculture, because he demonstrates

---

[6] Vernon W. Ruttan, "The Impact of Urban-Industrial Development on Agriculture in the Tennessee Valley and the Southeast," *Journal of Farm Economics,* Vol. 37, No. 1, February 1955, pp. 38-56. Reprinted in American Economic Association, *Readings in the Economics of Agriculture,* cited above, pp. 339-358.

that the patterns described above are true for both white and non-white farmers and farm communities.

Although Ruttan's work provides a convincing affirmation of some of Schultz's ideas, it has several shortcomings, which include its lack of historical perspective, its measure of a county's level of urban industrialization (he used percent of non-farm population, while Nicholls' work discussed below used levels of per-capita industrial value-added), and its weak investigation of the causal interrelations.

The results of nearly a decade of research in the 1950's by William Nicholls[7] document the inter-county relationships in the Tennessee Valley for the century from 1850 to 1950 and provide a fuller confirmation of Schultz's hypotheses. At the same time, a similarly comprehensive study of farm community income differences was carried out by Anthony B. Tang for the South Carolina-Georgia Piedmont.[8] Both studies showed that during the years 1850 to 1900, industry in the respective areas grew much more slowly than industry in other parts of the United States, and that during this period the equilibrating influence of migration reduced whatever discrepencies in farm community income had existed at the outset. By contrast, growth in both regions from 1900 to 1950 was well above the United States average, and farm community income inequality increased rapidly and in a pattern highly correlated with the regional development of urban-industrial centers. The studies showed cropping pattern changes as well, with advanced counties abandoning tobacco and cotton to a certain degree in favor of such products as livestock and poultry.

Nicholls and Tang looked mainly at markets for capital and labor in their efforts to explain the mechanisms of this correlation. The correlation of the farm capital-labor ratio with industrialization was very strong by 1954, though it had been insignificant at the turn of the century. Furthermore, study of the local capital market in industrialized centers showed a very large absolute increase in the size of bank deposits relative to bank deposits in isolated regions, indicating a much greater supply of capital available to the well-located farmer. Taking labor migration as an indication of the effectiveness of a region's labor market, Nicholls and Tang show that while migration to areas with more employment was sufficient to maintain

---

[7] William H. Nicholls, "Industrialization, Factor Markets, and Agricultural Development," *Journal of Political Economy*, Vol. 69, No. 4, August 1961, pp. 319-340.

[8] Anthony M. Tang, *Economic Development in the Southern Piedmont, 1860-1950: Its Impact on Agriculture,* Chapel Hill: University of North Carolina Press, 1958.

equilibrium in community incomes before 1900, such was not the case after 1900, when outmigration rates from poorer areas were not high enough to offset the much more rapid increase in value productivity for land nearer the industrial centers. Furthermore, their studies of secular trends in the age and sex structure of the corresponding labor forces confirmed the role of both out- and in-migration for countering, albeit unsuccessfully, the dominant trends. In their own words, one of the strongest findings of the two studies is that market adjustments and outmigration are not sufficient to maintain the relative level of a farm community's income if the community remains isolated from urban industrialization.

The studies referred to above have dealt with the Schultz hypotheses in the context of American agriculture. A second study by Nicholls,[9] based on the agriculture of the Sao Paulo region of Brazil, represents studies of this nature for a developing country and reflects a comparatively weaker supply of data.

Nicholls' study of 23 zones in the Province of Sao Paulo from 1940 to 1950 revealed that zones with higher levels of urban industrialization had higher farm incomes, higher per hectare value output, higher levels of milk production, a higher per hectare expenditure on chemical fertilizers, more investment in land, buildings, livestock and machinery, and better rural housing. Nicholls refers to the higher level of per-capita bank loans in industrial areas and the higher male share of the manufacturing workforce to bolster Schultz's hypothesis that capital and labor markets are more effective in those zones, but the results are not convincing enough to establish a causal relationship. In short, the results claim to confirm the phenomenon of differential farm community incomes but are unable to convincingly isolate the mechanisms responsible.

Nor is it clear that even the principal Schultz hypothesis has been substantiated, given the historical background of the Sao Paulo region. By Nicholls' own review of the region's history, the city and industrial areas of Sao Paulo owe their location to the initial productivity of agriculture in those areas. Furthermore, portions of Sao Paulo were still frontier regions undergoing active development in the decade of Nicholls' study, and hence it is not clear that the correla-

[9] William H. Nicholls, "The Transformation of Agriculture in a Semi-Industrialized Country: The Case of Brazil," in Erik Thorbecke (ed.), *The Role of Agriculture in Economic Development,* New York: National Bureau of Economic Research, 1969.

tions referred to are the result of an equilibrium in farm income patterns exogenously disturbed by localized industrialization.

South Korea, on the other hand, offers much better conditions for testing Schultz's propositions in a developing economy, given the centuries of intense and pervasive farm activity and given the pattern of initial regional equality of farm income documented by the provincial time-series statistics. Subsequent chapters do indeed document the following general statements in the context of South Korea's twentieth-century experience: (a) average incomes for farm communities are more equal before and at the beginning of industrialization than they are after industrialization in the economy has proceeded; (b) an economy's urban industrial development occurs within a spatial or locational matrix; and (c) the pattern of emerging inequality between farm communities reflects the locational matrix of the overall economy's development.

In all, the present study has seven additional chapters, a conclusion and three statistical appendixes. Chapter II introduces the economy of Korea from its opening to Japan and the West in 1876 to the end of the current study in 1975. Chapter III gives basic information about agriculture in South Korea, both in an international context and with emphasis on trends in production and overall distribution of produce, both between regions and within regions. Chapter IV is a brief introduction to the regions of South Korea and their differentiating characteristics, while Chapter V is a detailed analysis of regional patterns of agricultural growth and divergence in the period of Japanese colonial occupation (1905-1945). Chapter VI presents similar and in many ways more complete coverage for the period following World War II (1945-1975), and Chapter VII describes roads and other modes of transportation and communication, access to cities, and a method for the quantitative measurement of each individual county's degree of isolation from South Korea's industrial centers in 1970. Finally, Chapter VIII presents an econometric analysis of production function relationships using both pooled provincial cross-section and time-series and the purely cross-sectional data for 1970. The appendixes provide only a sampling of the provincial and county data series which support the analysis of the study's text.

The chapters which follow, then, marshal evidence from the South Korea experience which supports the findings and generalizations of T. W. Schultz. There are many shortcomings. Most noticeable is the lack of thorough research concerning the mechanisms responsible for urban influence on Korean agriculture. On the other hand, given the dearth of this kind of regional and historical analysis for rural Korea, and indeed for rural regions anywhere in the developing world, its findings may be both interesting and instructive for students of Korea as well as for students of rural development and economic growth.

CHAPTER II

KOREA AND THE KOREAN ECONOMY, 1876-1975

The more one researches agriculture in Korea, the more one realizes the importance of its relationship with the Korean economy as a whole, and in the century since its forceful opening by Japan in 1876, the economy of Korea has indeed experienced epochal change. From an initial condition of isolated equilibrium, the economy passed through colonization, political division, war and social turmoil, all of which left indelible marks on the overall pattern of change. Because of the importance of these developments for agriculture, they are reviewed briefly below. Both the structural patterns and the regional concentrations of Korea's general economic growth will be reflected in important ways in the subsequent detailed and analytical account of Korean agriculture's regional growth.

Political and international events have had such a strong influence on the Korean economy over the course of its first modern century that it is worthwhile summarizing them here. In 1876 the Japanese succeeded through force of arms in obtaining a treaty with the Korean kingdom opening the port of Pusan for trade. The opening of other ports and the expansion of international relations followed rapidly. Wars with China (1895) and Russia (1904) left Japan in a sole position of influence on the peninsula, and she proceeded first to establish a protectorate over Korea (1905) and then to annex Korea outright (1910), making the territory an integral part of the Japanese nation.

Popular uprisings (1919 and 1926) brought changes in the style of colonial control, while world depression (1929), Japanese military expansion in Manchuria (1931), and war in China (1936) and in the Pacific (1941) brought shifts in the nature of economic demands made on Korea by the parent Japanese economy. Throughout this period of colonial control, Korea's economic development was heavily integrated with that of Japan, and both investment in and regulation of the economy were explicitly in support of the 'home country's' economic and military goals.

Defeat in 1945 brought an end to Japanese control and Japanese ownership of assets, but the temporary United States occupying government (1945-48) was an ineffective economic administrator, and following the establishment of a Korean government (1948), there was time only for a land reform (1949) before the Korean War (1950-53) destroyed much of the south's industrial capacity and disrupted the population in ways which required many years of recovery. The influence of the United States on education and values in Korea was nevertheless very great, and throughout the Rhee government of the 1950's American influence in business and politics accumulated. A popular uprising (1960) and a military coup (1961) brought leadership change in most parts of the society, and a subsequent aid-independent program of very successful export-based industrial expansion has been marked by a peace treaty with Japan (1965) and the constitutional strengthening of President Park Chung Hee's executive power in the early 1970's. Buffeted by these events and bending to their pressures, the economy of Korea has thus been formed through various stages, each one taking it further away from its closed and feudal Yi Dynasty origins.

Korea's traditional 19th century economy was almost wholly dependent on agriculture and on a complex system of castes and servant-master relationships. Early Japanese economic historians and even some Koreans[1] have maintained that Korea's traditional economy offered little if any foundation for modern systems of economic activity, but this is certainly an exaggeration. In addition to a tradition of literacy and education and an excellent phonetic system for writing the vernacular, Yi Dynasty Korea had urban guilds for government and private handicrafts and a system of disciplined

---

[1] Much of the following account is based on Ho-chin Choi, *The Economic History of Korea,* Seoul: The Freedom Library, 1971; The Bank of Chosun, *Economic History of Chosun,* Seoul: Bank of Chosun, 1921; and Daniel Sungil Juhn, *Entrepreneurship in an Undeveloped Economy: The Case of Korea, 1890-1940,* Ph. D Dissertation, D.B.A., George Washington University, 1965.

itinerant peddlers who kept double-entry books and serviced the periodic rural markets held throughout the countryside.

The same writers note Korea's lack of economic sophistication when faced with the western powers and Japan and place the blame on Confucian values vilifying merchant activities, but this is probably only true insofar as government inactivity, corruption and open hostility to trade and industry placed obstacles in the way of popular commercial instincts which were themselves uninhibited by philosophical considerations. A brief further look at Yi Dynasty economic conditions will help place Japanese policies and accomplishments in perspective and will document the hypothesis that Korea's rural economy had reached a state of regional equilibrium by the late 19th century.

Yi Dynasty population records show that the country's total population was rather stable from the late 17th century through 1900.[2] Although Japanese statistics indicate Yi Dynasty figures underreported the population size, there is little reason to believe that underreporting was more severe in 1900 than in the 17th century. In addition, beginning in the 17th century there were signs that population had reached the level where it was putting pressure on the land's ability to support it with food. Examples of such signs are the beginning of official prayers for good harvests and government concern with irrigation and reservoirs to increase output.

Handicrafts formed the only manufacturing activities, and although the bulk of this work was done in farmers' homes, there were also government monopoly handicrafts using both freemen and indentured or bonded servants, as well as private handicraft establishments licensed by the government to provide the aristocracy with its needs: silk cloth and garments, paper, porcelains and pottery. There are accounts of guilds and official monopoly markets as well as a putting-out system for iron works and other products,[3] but serious studies of these and other aspects of the Yi Dynasty economy are difficult to find.

One interesting Yi Dynasty institution was the national organization of itinerant peddlers with a network of communications and discipline extending to periodic markets in all parts of the country.[4]

---

[2] See Juhn, cited above, pp. 25ff. for a more detailed discussion of Yi Dynasty population and reporting biases.

[3] See Choi, cited above, pp. 170-171.

[4] The account of the Kaesong peddler organization is from Juhn, cited above, pp. 42-46.

Reportedly formed from disenfranchised scholar officials after the fall of the Koryo Dynasty, the peddlers disciplined themselves to "fair and just" dealings on pain of expulsion from the organization, and also disciplined local officials for extortionary or other unfair treatment.

These peddlers provided the communications and transportation network for the more than a thousand rural periodic markets where foods and implements were exchanged, occasionally for money, but apparently almost always for each other. The peddlers' principal means of exchange were rice and cotton cloth, and a detailed Japanese history of currency use in Korea documents the complete lack of any uniform monetary system where money was in use at all.[5] This monetary picture was made chaotic after 1876 by the introduction of Japanese, Mexican, Russian and other national currencies, but there is some question as to the degree even these penetrated rural areas away from the few port centers of international activity.

A principal reason for dependence on peddlers who brought goods on their backs and for the rural insulation characterized by the widespread survival of pure barter was the apparent desperate condition of internal communications and transportation facilities:

> In the case of such thoroughfares as those between Seoul and Fusan [Pusan], Seoul and Chemulpo [Incheon] and Seoul and Wiju, the road was barely wide enough for vehicular traffic, and apart from these there were hardly any roads more than six feet wide and they, moreover, were exceedingly rough and barely fit for the passage of coolies and horses. Furthermore, many rivers had few, if any, bridges across them so that quite often, when ferryboats were not procurable, travellers were obliged to wade through the stream, and in times of flood were even obliged to stop for several days until the river again became fordable.[6]

The poor condition of these public facilities in turn reflected the reportedly pathetic administrative and fiscal capacity of the Yi Dynasty government. The personal financial perquisites of the Imperial household were intertwined with the responsibilities of the government's Financial Department.[7] There were no budgets, taxation was based on centuries-old records of land holding sizes and productivity, and rural rebellions were exemplified by but not limited to

---

[5] See Bank of Chosun, cited above, pp. 43-57.
[6] *Ibid.,* pp. 101-102.
[7] *Ibid.,* pp. 37ff.

the famous Tonghak uprisings. It is not possible to give here a detailed account of economic conditions during the last quarter of the 19th century, and we can only say that in spite of some official experimentation with electricity, trams, and railways, there seems to have been no effective Yi Dynasty program supporting the fundamental administrative and structural corrections necessary for a serious modernization effort. The slow pace of change in domestic conditions and the rapid march of international events, however, quickly left the job of reform and economic development in the hands of Japanese colonial administrators.

If the first thirty years following the opening of Pusan were characterized by the effective absence of government activity in support of the economy, the reverse was true for the remaining 70 years of Korea's first modern century. With the possible exception of the years between 1945 and 1961, the twentieth century development of Korea's economy has been the result of concerted and coordinated government policies and programs, all in cooperation with private economic forces which made a close alliance with government. These policies and programs either directly involved agriculture or had an indirect influence on agricultural development, and so it is worthwhile to review briefly the major trends and results of the seven decades of government economic policies in Korea since 1905. As the goals of the various governments have changed, so have their policies and successes varied, but they and the political events outlined above provide the necessary minimum basis for understanding the economy's fundamental patterns of progress.

The years of Japanese colonial control can be divided into four principal periods reflecting changing policies towards economic development and exploitation.[8] The first period, from 1905 to roughly 1919, was a period of preparation and development of infrastructure. The second from 1919 to the Manchurian Incident of 1931 was a period devoted largely to increasing agriculture output, in particular that of rice, and to securing as large a portion of Korean harvests as possible for export to Japan. The third period lasted only six years and represented a stage of "spectacular" industrialization on the peninsula.[9] Finally, the fourth period, from the 1937 start of war with China to the end of the War in the Pacific, was one of frenetic effort

---

[8] Takeo Suzuki, *Chosen no Keizai* (Economy of Chosun), Tokyo: Nihon Hyoronsha, 1942, (in Japanese), pp. 82-105.

[9] *Ibid.,* p. 91.

on the part of the Japanese to coordinate agriculture, mining and industrial activity in Korea in support of their military adventures. Changes for the Korean economy during these four decades were striking, and many of the institutions and institutional patterns established in those years, such as the Bank of Chosun, the system of secondary school education, the Ministry of Home Affairs, the secret police, the Agricultural Bank, and the complex of experimental farms and agricultural extension service facilities at Suweon, survived the colonial period almost intact and continued to function, if with perhaps varying degrees of influence, into the 1970's.

Of the four periods, the first was perhaps the most significant for Korea's twentieth century economy because of its sweeping reforms in all areas of Korean life and because of the institutional and statistical foundation it laid for all of the subsequent developments.

The construction of railroads had begun before 1905, the year in which the Seoul-Pusan line was opened, and a line from Seoul to the port of Incheon had been finished in 1900. By 1918 the Government Engineering Department had constructed 8,200 kilometers of highways, and postal, telegraph and telephone services reached all major cities in the country.[10] An extensive program of harbor improvement and warehouse construction began in 1906, and the consolidation of all coastal shipping under the Chosun Mail Steamship Company was completed in 1912.[11]

Other fundamental measures included monetary, banking and fiscal reforms and even a three-year survey of potential sites for hydroelectric power plants, but by far the most important event for agriculture and indeed for the economy as a whole at that time was the eight-year cadastral survey of all land in Korea. Begun in 1910 and completed in 1918, the survey established private ownership over all farmland for the first time, and in addition to providing a basis for land taxes, legalized and formalized the positions of landlords and tenants. Through their understanding of this survey, the Japanese-owned Oriental Development Company (founded in 1908) and private Japanese citizens dramatically increased their ownership of and control over Korean farmland.[12] Undertaken at great expense and

---

[10] Bank of Chosun, cited above, pp. 102-103.

[11] *Ibid.,* pp. 107-109.

[12] For an historical account of this and other aspects of Japanese ownership of Korean land see "Chosun Ilbonin Tochi Soyu ui Teoksu Gwa Ke Cheobun Munje" [Characteristics of Former Japanese Land Ownership], *Monthly Statistical Review,*

employing thousands of Japanese and Korean workers, the survey allowed more accurate crop reporting and, in addition to providing a foundation for landlord control and the collection of grain for export, also formed the basis for the Rice Production Increase Plan and the Farmland Improvement Plan of the 1920's

In sum, this first economic period was one of dramatic investment in all parts of the new Korean economy's institutional and statistical base. The impact of the period's major international event, World War I, was felt only indirectly through the increase in demand from Japan for processed Korean foodstuffs, as Japan's own manufacturing industries expanded and profited from the war.[13] The first period also saw an efficient and naked establishment of civil order and political obedience through the use of a Japanese Gendarmerie, and it is interesting that the end of World War I and Wilson's 14 points inspired Korean demonstrations in March 1919 which, along with the over-reaction of Japanese police, forced a change in style to one of more indirect and subtle control which was, however, no less effective in maintaining domestic discipline.

The goals of the Japanese Government during the second period (1919-1931) were summarized in the part of the 1921 *Report of the Chosen Industrial Survey* entitled "The General Plan for Chosen Industry."[14] The industry of Korea (in which was included agriculture) was to be coordinated with the needs of the "home country," and in particular the increase in rice output, the decrease in Korean domestic rice consumption and the substituting import of Manchurian millet were all meant to support the export of rice to Japan.[15] Study shows that increases in rice imports to Japan from colonial Taiwan and Korea either were needed because of, or perhaps made possible, a slowing down in the rate of productivity increase in Japan's agriculture production and hence freed the Japanese government from the necessity of spending additional large sums on further domestic yield increases.[16]

The long-run policy for increased rice exports was set out in the

Bank of Chosun, No. 15, August and September 1948, (in Korean), pp. 125-135.

[13] Suzuki, cited above, p. 85.

[14] For excerpts, see Suzuki, cited above, pp. 86-87.

[15] *Ibid.*, pp. 87-89.

[16] Yhi-min Ho, "Korean Rice, Taiwan Rice, and Japanese Agricultural Stagnation: An Economic Consequence of Colonialism," Paper No. 16, Program of Development Studies, Rice University, Summer 1971.

1918 "Plan for Increased Rice Production,"[17] which was to be a fifteen-year program of measures to increase output mainly by improving land, seeds and techniques. The Great Depression, falling world agricultural prices and rice riots by Japanese farmers forced an abandonment of the plan in 1931,[18] but part of the plan was successfully carried out, and as later chapters will show, there were significant increases in yields in some parts of the country.

Other non-agricultural policies of this 1919-31 period resulted in the 1926 formation of the Chosun Hydroelectric Power Company, the 1927 opening of the Chosun Chemical Fertilizer Company and the first transmission of power from a Korean hydroelectric plant in 1929.[19] In spite of the potential for mining a variety of minerals in North Korea, mining in this period was limited to gold, silver and smokeless coal for Japan's navy.[20]

The Depression and rice riots mentioned above on the one hand, and the military developments following the Manchurian incident of 1931 on the other brought about a major change in this basic Japanese policy for Korea. The contraction in world-wide trade and the pressures on foreign exchange made Korea's mineral resources suddenly very valuable to Japan. In addition, the development of the "North Korean Mainland Route" notions of access to Manchuria and China by way of Korea gave a prominence to industry in Korea which had previously been lacking.[21] The degree and nature of the resulting industrial expansion in the third (1931-36) period can be seen from Table II-1, "Korean Industrial Employment Structure, 1930-39." The most dramatic increases in industrial employment between 1930 and 1936 were in chemicals, machinery, lumber and textiles, but in absolute numbers, chemicals and textiles were by far the most important. It is interesting that the traditionally significant food processing industry experienced a much smaller expansion than many of the other sectors. In short, the end of the rice production increase

---

[17] For a more detailed history of the rice production increase program see Chapter 3, "Regional Agricultural Development in the Japanese Period, 1910-1945" below. See also Kuro Komikawa (ed.), *Chosen Nogyo Hattatsu-shi, Seisaku-bu* [History of Chosun Agricultural Development, Policy Volume], Tokyo: Chosen Nogyo-sha, 1944, (in Japanese). pp. 418ff., and Choji Hishimoto, *Chosen Mai no Kenkyu* [A Study of Chosun Rice], Tokyo: Chikura Shobo, 1938, (in Japanese), pp. 52ff.

[18] See Hishimoto, cited above, pp. 71-73.

[19] Suzuki, cited above, pp. 90-91.

[20] Choi, cited above, pp. 223-226.

[21] Suzuki, cited above, pp. 91-94.

Table II-1. Korean Industrial Employment Structure, 1930-1939

	1930		1936		1936[a]		1939[a]		1930-36 annual growth (%)	1936-39 annual growth (%)
	Workers	Share	Workers	Share	Workers	Share	Workers	Shares		
Textiles	19,001	(22.6)	33,830	(22.7)	36,520	(19.4)	52,081	(19.3)	10.1	12.6
Metals	4,542	(5.4)	6,787	(4.6)	7,874	(4.2)	17,875	(6.6)	6.9	31.4
Machinery	2,854	(3.4)	7,939	(5.3)	9,065	(4.8)	29,579	(10.9)	18.6	48.3
Ceramics	5,366	(6.4)	8,269	(5.6)	11,098	(5.9)	15,162	(5.6)	7.5	11.0
Chemicals	14,720	(17.5)	41,972	(28.2)	54,845	(29.1)	71,673	(26.5)	19.1	9.3
Lumber	2,629	(3.1)	4,906	(3.3)	7,268	(3.9)	12,401	(4.6)	11.0	19.5
Printing	4,146	(4.9)	6,273	(4.2)	7,843	(4.2)	8,403	(3.1)	7.1	2.3
Food Processing	27,055	(32.2)	32,617	(21.9)	46,496	(24.7)	48,610	(18.0)	3.2	1.5
Gas and Electricity	525	(.6)	812	(.5)	1,232	(.7)	1,336	(.5)	7.5	2.7
Other	3,052	(3.6)	5,394	(3.6)	6,009	(3.2)	13,319	(4.9)	10.1	30.4
Total	83,900	(100.0)	148,799	(100.0)	188,250	(100.0)	270,439	(100.0)	10.0	12.8

Source: Takeo Suzuki, Chosen no Keizai [Economy of Chosun], Tokyo: Nihon Hyoron-sha, 1942, (in Japanese), pp. 223, 233.
Note: [a] These 1936-1939 data include employment in smaller factories excluded from the 1930-1936 data and hence are not directly comparable. Examination of the two series for 1936, however, reveals their rough equivalence.

plan was offset by an emphasis on investment in non-traditional manufacturing in the first six years of the 1930's.

In addition, the depression years were not easy for farmers, and in particular for tenant farmers. Japanese statistics show that the numbers of rural people searching through the woods and fields for edible bark, grasses and nuts during the period of 'spring hunger' before winter barley harvests passed 50 percent of the rural population, and for landless tenant farmers the incidence of spring hunger was greater than 70 percent in the southern half of the country.[22] The decade of the 1930's also saw substantial migration of Koreans to Japan and Manchuria, much of it originating in the southern agricultural provinces.[23] This migration, which exceeded 400,000 persons between 1930 and 1935, increased to a net outflow of over a million between 1935 and 1940. In this respect and in others, the difference between the third and fourth periods of Japanese colonial development in Korea seems to be more one of degree than substance, but the placing of the entire Imperial economy in a war-preparatory state also had its effect on policy in Korea, and it is therefore helpful to separate the periods before and after 1937.[24]

As indicated by the level of out-country migration, the wartime years were difficult ones for the Korean economy. Further study of Table II-1, "Korean Industrial Employment Structure, 1930-1939," shows that trends in industrial expansion begun in the previous period became much more exaggerated. Expansion of metals, machinery and lumber was considerable in both percentage and absolute terms, while food processing barely grew at all. The change in industrial structure was paralleled by changes in the composition of minerals mined. Although gold remained important, iron, copper, tungsten, graphite and other industrial metals rapidly gained importance in the 1930's and early 1940's.[25] The period also saw a shift in agriculture towards the production of cotton and wool, while increased forced taxation in kind of foodgrain production became severe after 1941.

To summarize, Korea's integration into Japan's wartime economy

---

[22] See the table in Choi, cited above, p. 268; the data are from Chosun Government General, *Chosen ni okeru Kosaku ni kansuru Sanko Jiko Tekigo* [Reference Materials for Tenancy in Chosun], 1932. (in Japanese)

[23] Figures for migration in this and other periods are from Tai-hwan Kwon, *The Population of Korea*, Seoul: The Population and Development Studies Center, Seoul National University, 1975, p. 29.

[24] For a development of this reasoning see Suzuki, cited above, pp. 97-105.

[25] Choi, cited above, pp. 277-279.

was the fourth and final stage in a process of economic development which began with a wholly agricultural nation devoid of modern economic institutions and ended with an economy whose manufacturing and mining capacities and whose transportation and communication networks were vital components in the economy of imperial Japan. The degree of Japan's dependence on Korea can be seen by noting that Korea in 1940 had a population over a third that of Japan itself and more than four times as large as that of Japan's next largest colony, Taiwan.[26] Furthermore, in terms of trade, Korea in 1939 was Japan's largest foreign market and accounted for over 23 percent of all Japanese exports. In return she provided over 17 percent of Japan's imports, second only to the United States.[27] The transformation of Korea's economy within the space of 40 years had been rapid, but it had been also painful for the Korean population, and many difficult years would be required to correct the imbalances imposed during those four decades.

The 1945 defeat of Japan and the liberation of Korea into the hands of her Russian and American benefactors brought at once an end to the strongarm colonial control of Korea's economy and for the South the beginning of a long period of relative chaos and uncoordinated economic activity. After political division, bitter civil war and slow recovery, it was not until 1961 that a strong government once again began to make its will felt on the Korean economy. Furthermore, just as agriculture took a place of secondary importance for colonial economic policy after 1930, so agriculture in the years following World War II was neglected in favor of manufacturing. The paragraphs which follow will outline briefly the well-documented pattern of Korean economic growth after 1945.

The post-1945 years can also be divided into periods or stages for convenient analysis, and the periods again reflect the influence of political or international events important for Korea's economy. The first period, from 1945 to the end of the Korean War in 1953, was one of turmoil, destruction and dislocation, while the second period, from 1953 to 1961 was one of recovery from war, of government failure and of political change. The year 1961 was a turning point of perhaps singular importance, because it marked the assumption of power by a government both willing and able actively to encourage

---

[26] Suzuki, cited above, p. 22.
[27] *Ibid.*, pp. 294-295.

economic development. The years from 1961 to 1966 are significant as preparatory years for the spectacular growth which was to follow and coincided with the span of the government's first five-year plan for economic development. It is also significant that the period's last years saw the signing of a peace treaty with Japan and the resumption of large-scale economic intercourse between the two close neighbors.

The fourth post-1945 period followed the normalization of relations with Japan and extended to the "oil shock" of 1973. It was during these years, which also coincided with the United States' greatest military and economic presence in Vietnam, that Korea experienced the very strong and sustained overall economic growth which placed her among the important trading nations of the world and marked her as an example of "success" in the process of economic development. Finally, the period from 1973 to 1975 and beyond is one of unexpected rapid recovery from the effects of explosive energy costs and one in which the strategy of development in Korea had begun to change in terms of markets and the structure of industrial expansion. In sum, the last 30 years of Korea's modern century have seen stages of change perhaps as dramatic as those under Japanese colonial control, and although the emphasis has been on manufacturing throughout most of the period, subsequent chapters will show that the influences on agriculture have also been striking.

The 1945 division of Korea into northern and southern zones left much of the nation's industrial capacity in the North, and the much greater part of its agricultural resources in the South. As unbalanced and dependent on Japan as the late-colonial Korean economy had been, it had, in fact, a large measure of integrity when compared to the fractured and distorted structure of South Korea's 1945 economy. Unable to provide its own manufacturing wants, exhilirating changes in food consumption patterns deprived South Korea of even the expected surplus in foodgrains, and as a result the economy found itself heavily dependent on imports of consumer goods from the United States for the satisfaction of its basic wants. Neither the interim United States military government nor the civilian government which took its place saw the peninsula's division as permanent, and hence there was no serious attempt to formulate ways to correct the zone's fundamental economic imbalances.

These governments did, however, carry through with perhaps the single most significant economic policy of the entire post-liberation era when they redistributed farmland to farmers in the 1949 land reform, converting Korean agriculture from a feudal tenant-bound industry to one of owner-tiller husbandry. This sweeping albeit imperfect re-allotment of assets created a climate of equality by any internationally comparative standards, and it was this equality in human potential and aspirations which, when later combined with education, transportation and urban opportunities, possibly formed the richest fuel for sustained and pervasive economic combustion.

Aside from the land reform and the further leveling influence of the Korean War's almost complete destruction of Seoul's industrial capacity, the 1945-53 period left few tangible marks on the South Korean economy. Statistics for subsequent periods, however, provide a basis for comparison and measures of the successes and their components. Table II-2, "National Economic Growth Statistics, 1953-1975," shows that the seven-year period from 1966 to 1973 experienced an average rate of 10.7 percent annual growth in gross national product, an extremely high rate to be sustained for so many years. Manufacturing during this same period grew at an average rate of 23.3 percent, and since that sector accounted for over 22 percent of gross national product for those years, its contribution to the overall growth rate of the economy is unmistakable.

The same table indicates further reasons for the difference in the growth record between the 1966-73 period and the years of the 1950's. Gross investment in the high-growth period exceeded a fifth of gross national product and was financed to a large degree by a combination of domestic savings and foreign borrowing. This pattern of investment finance contrasted sharply with that of the pre-1961 period, when foreign transfers, that is to say, aid grants, accounted for more than 80 percent of investment. In sum, these statistics indicate that the most successful years of the latter 1960's and early 1970's resulted from economic activity of much more vigor and integrity than that of the 1950's. The reasons for these differences lie in the economic policy strategy of the Park government and its contrast with that of the earlier Rhee regime.

Put succinctly, the post-1961 policies stressed exports, planning and the resumption of relations with Japan. Where the Rhee government had sought foreign aid to pay for consumption imports, the Park government sought capital, raw materials and foreign markets for the products of the economy's inexpensive yet highly trainable labor force. Typical of the shift in personnel responsible for policy from

Table II-2. *National Economic Growth Statistics, 1953–1975*

(%; based on 1970 constant prices)

	Annual GNP Growth	Annual Manu- facturing Growth	Gross Investment ÷ GNP	National Saving ÷ Gr. Inv.	Net Foreign Transfers ÷ Gr. Inv.	Net Foreign Borrowing ÷ Gr. Inv.	Statistical Discrepancy
1953-61	4.4	10.7	11.1	23.3	80.1	−5.7	−2.3
1961-66	7.6	13.5	14.7	38.4	49.2	6.8	−5.6
1966-73	10.7	23.3	21.9	61.4	13.0	22.1	−3.5
1973-75	8.0	24.4	25.7	70.7	4.8	28.6	4.1

Sources: Calculated from Bank of Korea, *Economic Statistics Yearbook*, Seoul, 1974, 1976.

Japanese-era bureaucrats to post-liberation college graduates was the eclipse of the Bank of Korea and its economic research department by the new and assertive Economic Planning Board, which during the course of the 1960's accumulated influence and discretion until it obtained a position of pre-eminence in matters of economic regulation and initiative. The govenment's policies were coordinated in a series of five-year plans which provided a framework for investment and market expansion. In addition, the government actively supported key firms and industries and provided a combination of incentives, subsidies and monopoly rights to ensure the financial success of new or growing ventures in strategic sectors.

The rapid industrialization of the post-1961 era, in summary, resulted from political forces which took advantage of two sets of circumstances, one domestic and the other international. The domestic circumstances were based on the presence of a new generation of potential laborers from an egalitarian society, providing a universal minimum standard of education on the one hand and the evolution of a new generation of managers and technicians capable of filling the void left by repatriated Japanese personnel on the other. The international circumstances involved access to adequate American and Japanese markets for manufactured exports and the willingness of Japanese and American capital to invest in plant and equipment in Korea. Planning, incentives and government discipline proved sufficient catalytic agents for bringing these forces together in successful combinations, and the results have been growth and diversification perhaps as rapid as the bounds of social stability can permit.

The twentieth century growth of Korean industry, then, has spanned decades of colonial control, periods of war and political chaos and an era of planned and internationally coordinated economic expansion. The significance of these events for agriculture and the rural economy of Korea has been of two kinds. First, as industry has expanded in the twentieth century, the importance of agriculture in the economy's value output and share of labor force has drastically declined. Second, the industrial expansion has not been geographically uniform, and the spatial pattern of modern growth has left in turn its spatial influence on growth and diversification of farm production and income.

The severity of the decline in agriculture's relative significance between 1910 and 1975 can be seen by looking at Table II-3, "Shares of Agriculture in Total Output and Population, 1910-1975." The

Table II-3. Shares of Agriculture in Total Output and Population, 1910-1975

(%)

Years	Farm Product as a Share of National Output[a] (i)	Farm Population as a Share of Total Population (ii)
1910-12	84.6	83.0[b]
1919-21	78.6	—
1929-31	63.1	78.4
1939-41	49.6	—
1949	—	71.1
1955	44.9	61.8
1960	39.9	58.3
1966	36.7	54.1
1970	24.5	45.9
1975	17.9	35.3

Sources: Column (i), 1910-1941 from Sang-chul Suh, "Growth and Structural Changes in the Korean Economy Since 1910," Statistical Appendix, Table A-1. Column (i), 1955-75 from output data, Appendix C and GNP data, Bank of Korea, *Economic Statistics Yearbook,* 1974, 1976.

Notes: [a] 1910-41 data are from current price calculations; the remainder are from calculations using constant 1970 prices. It should be noted that the data for 1910-12 are of limited reliability. For a discussion see Sung-hwan Ban, "The Long-Run Productivity Growth in Korean Agricultural Development, 1910-1918", Ph. D Dissertation, 1971, University of Minnesota, Chapter 2.

[b] Based on farm households and total households rather than on population.

share of Korea's population engaged in agriculture fell from 83.0 percent in 1910 to 35.3 percent in 1975, while the fall in share of value output was even more drastic, from 84.6 percent of the total to 17.8 percent, reflecting also a decreasing relative value of output per member of the farm population compared to per capita output in the non-farm sectors. As we shall see in subsequent chapters, the incidence of these declines has not been uniform throughout the regions of Korea, and the rapidity of decline is in some ways related to the relative differences in per capita farm value output.

The second important aspect of industrial growth for agriculture is its geographical variation. Industrial urban areas appear in some parts of a country but not in others. The paragraphs and tables which

Table II-4. *Manufacturing Population in North and South Korea,*[a] *1915-1943*

(thousand persons)

Year	North Korea			South Korea			Total	Average Annual Growth (%)
	Manu-facturing Population	Share (%)	Average Annual Growth (%)	Manu-facturing Population	Share (%)	Average Annual Growth (%)		
1915	70.3	(31.5)	—	152.8	(68.5)	—	223.1	—
1927	157.3	(32.5)	6.9	326.3	(67.5)	6.5	483.6	6.7
1930	173.7	(32.7)	3.4	357.8	(67.3)	3.1	531.6	3.2
1935	287.3	(45.5)	10.6	344.3	(54.5)	-.8	631.6	3.5
1941	476.3	(44.4)	8.8	595.3	(55.6)	9.6	1,071.7	9.2
1943	671.0	(45.3)	18.7	811.8	(54.7)	16.8	1,482.8	17.6

Sources: Calculated from data in: Chosun Government General, *Chosun Statistical Yearbook*, relevant years.
Note: [a] All of pre-1945 Kyonggi and Kangwon Provinces are included in South Korea for this table.

follow will briefly show the degree to which this has been true for twentieth-century Korea and subsequent chapters will document in detail the spatial pattern of change in agriculture.

Table II-4, "Manufacturing Population in North and South Korea, 1915-1943," confirms the geographical shift in the pattern of manufacturing development after 1930. While manufacturing employment actually contracted in the South between 1930 and 1935, it grew at an annual rate greater than 10 percent in the North, increasing the North's share in Korean manufacturing employment to 46 percent from its 33 percent level in 1930. Both before and after this five-year period, growth in the two halves of the country was relatively even, and it is interesting that after 1935, when the war preparation became intense, the growth in overall manufacturing employment dramatically accelerated to where it was growing at an annual rate of greater than 15 percent by the early 1940's.

More interesting for our purposes than the North-South pattern of growth is the more detailed geographical picture of manufacturing employment given by statistics for individual provinces. Table II-5, "Shares of Provincial Manufacturing Population, 1915-1943," shows the share of the manufacturing population (workers and their families) out of the total provincial population. For many provinces, particularly in the more agricultural South, the relative size of the population dependent on manufacturing increased very little over the nearly thirty-year period covered. For other provinces, the shares of total population increased from less than 3 percent in all cases to between 9 and 14 percent.

Naturally, these provinces with the most rapid changes were those with industrial cities. In the North the three provinces with large shares of manufacturing population were those containing the cities of Pyongyang, Weonsan and the triple combination of Chungjin, Ranam and Kyeongseong. In the South, Kyonggi Province, which contains the city of Seoul, stands out. It is particularly important to remember this imbalance centered on Kyonggi Province in the South, because a later chapter will document how in fact the increase in Kyonggi's agriculture productivity also outpaced that of all other provinces in the South.

If the geographical pattern of industrial activity was skewed by the end of the Japanese colonial period, it became even more so during the 30 years following World War II. The differences from province to province in the relationship of manufacturing to agriculture are

Table II-5.   *Shares of Provincial Manufacturing Population,* [a] *1915-1943*

(%)

	1915	1927	1930	1935	1941	1943
North Korea						
Hwanghae	1.0	1.9	2.1	2.5	2.3	3.4
S. Pyongan	2.2	3.7	3.7	5.7	7.0	9.7
N. Pyongan	.7	2.2	2.3	3.2	2.9	3.5
S. Hamgyong	1.6	2.4	2.6	5.0	7.6	9.0
N. Hamgyong	.8	2.7	3.0	4.5	10.0	13.5
South Korea						
Kyonggi	2.6	5.0	5.2	4.8	9.2	11.7
Kangwon	1.0	1.5	1.4	1.8	1.9	2.3
N. Chungchong	.9	1.3	1.6	1.2	1.1	1.6
S. Chungchong	1.2	2.0	2.2	1.6	1.4	1.9
N. Jolla	1.3	2.5	2.5	1.8	1.8	2.3
S. Jolla	1.0	1.9	2.0	1.7	1.9	2.6
N. Kyongsang	1.1	2.2	2.4	1.8	3.0	3.5
S. Kyongsang	1.7	2.6	2.5	2.6	4.4	5.4

Sources: Chosun Government General, *Chosun Statistical Yearbook,* individual issues for relevant years.

Note: [a] Provincial manufacturing population is taken as a share of total population in each province.

made extremely clear by Table II-6, "Regional Manufacturing Value Added per Farmer, 1960-1974." Ignoring for the moment the very large ratios for Seoul and Pusan, there are three provinces with manufacturing value added per farmer significantly higher than that for other regions. Kyonggi and South Kyongsang envelop the two above-mentioned cities, and North Kyongsang contains South Korea's third largest industrial city, Taegu, well-known as a textile center. If we imagine that the cities have in addition a spill-over effect influencing surrounding regions to which there is easy access, it will be easier to understand the influence of urban centers on the geographical patterns of agricultural change which became evident after 1945.

Finally, whatever the effect of this concentration on agriculture, its considerable influence on overall per capita product is shown by Table II-7, "Regional per Capita Gross Domestic Product, 1955-1972." The three provinces referred to above have higher per-person

Table II-6.   Regional Manufacturing Value Added per Farmer, 1960-1974

(thousand 1970 won[a])

	1960	1963	1966	1970	1974
Seoul	533.0	411.8	633.9	2,571.5	5,842.4
Pusan	—	346.3	605.4	1,787.9	3,137.0
Kyonggi	4.7	8.0	11.9	39.9	70.6
Kangwon	3.5	2.8	6.0	6.5	11.2
N. Chungchong	1.4	6.4	8.5	18.1	16.0
S. Chungchong	2.2	2.7	6.0	21.1	18.2
N. Jolla	2.1	4.9	4.9	13.5	12.9
S. Jolla	1.2	2.3	3.6	5.7	16.3
N. Kyongsang	3.9	6.3	8.3	17.3	30.5
S. Kyongsang	7.8[b]	2.2	6.6	32.5	38.9
Jeju	.8	3.4	3.4	6.5	4.7

Sources: Manufacturing data from Economic Planning Board, *Korea Statistical Yearbook,* 1961, 1967, 1976. Farm population from Appendix B, Table B-1.

Notes: [a] The original manufacturing value added in current prices was deflated by the Bank of Korea's wholesale price index because no price index for manufacturing goods was available. Nevertheless, the price indexes for principal sectors of manufacturing (chemicals, machinery, ceramics, rubber products and others) move sufficiently in step with the index used here to ensure its suitability for the purposes of this table.

[b] The South Kyongsang data for 1960 include Pusan.

product than any of the other provinces but one, and we will see in the chapters on post-1945 agricultural growth that this fourth province found itself in particularly advantageous circumstances.

The present chapter has reviewed the economy of Korea during its first hundred years of exposure to foreign contact, with an emphasis on the growth of manufacturing and the significance of patterns in that growth for regional agricultural development. The century's changes were stunning in many ways, and it is against this background of the birth and build-up of modern capacity that we can perhaps best understand agriculture's own special path.

Table II-7. *Regional per Capita Gross Domestic Product, 1955-1972*

						(thousand 1970 won per capita)
	1955	1960	1962	1966	1970	1972
Seoul	44.0	62.3	58.7	—	—	176.3[a]
Pusan	—	—	—	68.6	118.6	125.1
Kyonggi	22.7	26.6	26.8	40.7	76.2	88.2
Kangwon	19.0	26.7	28.0	37.0	64.2	72.4
N. Chungchong	19.9	21.5	23.5	39.8	69.9	82.6
S. Chungchong	19.5	21.7	21.9	37.0	63.9	77.2
N. Jolla	21.2	21.8	23.0	36.7	61.3	79.6
S. Jolla	21.7	19.9	22.4	34.0	56.5	70.1
N. Kyongsang	22.9	21.7	25.2	37.8	61.6	80.7
S. Kyongsang	23.0	24.4	26.5	37.1	73.6	95.2
Jeju	17.0	25.0	28.2	42.6	65.7	80.6
South Korea	23.1	26.8	28.4	47.6	81.3	121.0

Source: Young-il Chung, "Over-time Changes in the Regional and Urban-rural Income Differences in Korea," Kunitachi, Japan: Institute of Economic Research, Hitotsubashi University, May 1976, (mimeographed), p. 19. The data were adjusted to 1970 levels by the Bank of Korea's wholesale price index. The original data are from: (1955) USOM-Korea estimates; (1960-62) Bank of Korea, Research Department; (1963-72) Ministry of Home Affairs; (1973) Korea Development Institute.

Note:[a] This column's data for Seoul are from 1973.

CHAPTER III

KOREAN AGRICULTURE

The agriculture of the Korean peninsula is in a sense representative of the overall East Asian agricultural pattern. Nevertheless, the general development of Korea's agriculture in the twentieth century is a unique story of output growth, shifting structure and changing distribution. These principal features are introduced briefly below and will become the focus of more detailed regional analysis in later chapters.

Korea's rural economy mirrors that of China and Japan in being rice-dependent and densely populated. In addition, just as China's agriculture specializes in barley and wheat in the North and rice in the more densely populated South, so Korea's southern rice belt region is much more heavily settled than the barley and wheat areas of what is now North Korea. Our description here will concentrate on the agriculture of South Korea, which by international standards is extremely densely populated and traditionally heavily dependent on the success or failure of a single annual rice crop.

Table III-1, "International Agricultural Statistics," shows that of the countries listed South Korea has a farm population density higher than any of the others, over seven thousand people per thousand hectares of arable land. This is of course in stark contrast to United States farm density, but is also considerably higher than even that for Taiwan, Indonesia and Japan. The same table shows also that rice yields in South Korea are higher than in any of the other developing economies, although not so high as in Japan and the United States. In these terms, then, South Korea is unique in the developing world for

Table III-1. International Agricultural Statistics

	Farm Population (1000s)	Farm Land (1000 hectares)	1973 Rice Output (1000 tons)	Rice Yields (kg/ha.)	Farm Density (Persons /1000 ha.)
Belgium	469	840	—	—	558
Columbia	9,652	5,044	1,050	3,804	1,913
India	364,823	165,580	67,600	1,827	2,203
Indonesia	84,839	18,100	20,321	2,372	4,695
Japan	21,564	5,296	15,766	6,018	4,065
S. Korea	17,132	2,241	4,212	3,560	7,634
Taiwan	5,868	896	2,452	3,114	6,579
United States	8,192	191,053	4,210	4,794	43

Source: Statistics from the F.A.O. 1973 Yearbook compiled in Republic of Korea, Ministry of Agriculture and Fisheries, *Statistical Yearbook,* 1975, pp. 434ff.

the close-packed density of its farm population and for the degree to which that farm population has been able to squeeze rice from soil which is scarce and less favored by weather than in other more tropical economies.

Given this fundamental imbalance in the man-land relationship, it is not surprising to find that Korean farm technology is very labor-intensive. Whether we speak of pre-land-reform tenancy or post-1950 owner-husbandry, Korean farming is basically family farming. Furthermore, although the introduction of engines and machinery has brought important changes for some parts of the production and preparation process, other parts remain heavily dependent on human effort, and in most areas draft cattle are the principal auxiliary source of power. The degree to which farming in Korea is a family business is seen in data on farm labor use.[1] Throughout the 1960's and 1970's family labor supplied between 70 percent and 80 percent of total farm labor needs, while hired labor (more heavily male in composition) supplied roughly 18 percent, the remainder being made up of exchange labor from other households. It is also interesting that the same data show a trend towards greater use of family and less use of hired labor.

---

[1] Ministry of Agriculture and Fisheries, *Report on the Results of the Farm House-hold Economy Survey*, Seoul, 1975, pp. 84-85.

The principal tasks in the Korean agricultural cycle for rice are seed preparation, seedbed planting, paddy repair, tillage, transplanting, irrigation maintenance, fertilizer and pesticide application, weeding, harvesting, threshing and polishing. Each of these is in turn composed of specialized interrelated tasks, and with the exception of tillage and transport, these have been done mostly by humans. Machinery has gradually replaced hand work in some activities, most notably in tilling, threshing, and polishing (the Japanese placed early emphasis on mechanizing and centralizing the polishing process), but by the mid-1970's almost all of the particularly grueling work for transplanting and harvest was still being done by hand. For crops other than rice the activity cycle is less complex and less drawn out, but it is nevertheless equally dependent on human labor and similarly seasonal in its pace.

We will not attempt here to give a detailed analysis of Korean farm technology, much less describe the advances it has made in the course of the twentieth century. It is sufficient to note that such progress has involved increases in quality and quantity of inputs (chemical as opposed to natural fertilizers, new seed varieties, irrigation and chemical pesticides) as well as changes in the technique and skill of labor. Some idea of the complexity of the human input and the subtlety of potential improvements is provided by the following passage describing rice techniques in Ganghwa County, Kyonggi Province in the late 1940's:

> In preparation for transplanting the rice, two special lines are stretched along the long sides of the field to be filled with seedlings. The field should have a little water on the surface, but water may be short. These lines have marks approximately eight inches apart. Then a cross line, handled by two boys or girls, is extended across the field, beginning with the eight-inch marks at one end. The men, as they move backwards down the paddy, can manage the string themselves if children are not available. The workers line up each holding part of a bundle of seedlings in his left hand, palm-up, with the roots extending beyond the little finger. Each man draws out three seedlings between the thumb and first two fingers of his right hand, inserting the roots in front of the line by using his second finger to punch a hole in the mud to half its length. The seedlings are placed about seven inches apart laterally, thus making almost a square of plants as the cross line is moved to the next eight-inch marker.
>
> The villagers learned this system of transplanting, together with the manner of soaking seeds and laying out seed beds in approachable units with better drainage, under Japanese tutelage in the regime of Governor

Ugaki, or during the decade between 1930 and 1940. The straight lines in the growing fields decrease damage during cultivation, both from walking on and turning the soil. The reduction to three seedlings from the previous five or six eliminates crowding, for seedlings double, and the new stalks bear most of the rice. Actually, many Koreans still plant five or six seedlings together, thus cutting down production by stunting the heavy bearing stalks. Near the end of the war, the custom of making a small furrow where the line crosses the field and of planting seedlings only two or three inches apart on the intervening ridges was recommended. This method allows more sun to strike the roots, thereby increasing the yield as much as one third. Also, cultivation is easier. Japanese instruction ended with the American occupation, by which time few farmers had learned the new procedure. In the old days, the people distributed their seedlings somewhat haphazardly.[2]

Next to harvest, rice transplanting is the most demanding activity, and the same basic process described above continued in almost universal use in the mid-1970's, though the appearance of a few highly publicized and for the ordinary farmer quite exotic power transplanters in the counties near Seoul is an indication of trends for the 1980's and beyond.

The rapid fall during the twentieth century of Korean agriculture's share in total output has provided a strong indication of the relatively slow pace of growth in overall farm product. But although growth in output has been slow, it has not been uniformly so. Table III-2, "South Korean Agricultural Growth Rates, 1920-1969," shows that the average annual growth rate of farm product was considerably higher after the Korean War than it was in the Japanese colonial years before World War II. Some remarks are appropriate about the two sets of growth rate estimates presented. Ban's estimates are the result of exhaustive efforts at a comprehensive measure of total farm product, value added and productivity sources for most of South Korea's twentieth century. Ban has been particularly resourceful in calculating estimates of livestock output, and the absence of livestock data in the figures used for this study's estimates up to 1953 accounts for the different results for the two estimates in this period. For the period after 1953, both estimates include livestock, and hence the results are more comparable, but differences exist and are due in large

---

[2] From an account of two years of anthropological field work in 1947 and 1948, in Cornelius Osgood, *The Koreans and Their Culture,* Rutland: Charles E. Tuttle, 1951, pp. 66-67.

Table III-2.   *South Korean Agricultural Growth Rates, 1920-1969*

(percent per year; based on five-year averages)

	Ban's Estimates (constant 1934 prices)		This Study's Estimates[a] (constant 1970 prices)	
1920-69	1.9		—	
1920-39	1.6		—	
1920-30		0.5		—
1930-39		2.9		—
1939-53	−0.3		0.5	
1939-45		−3.5		0.0
1945-53		2.1		0.9
1953-69	4.4		4.2	
1953-61		3.6		4.0
1961-69		5.1		4.5

Sources: Ban's estimates are from Sung-hwan Ban, "Growth Rates of Korean Agriculture (1918-1971)," in Sung-hwan Ban, *Hangug Nongeop ui Seongchang* [Growth of Korean Agriculture], Seoul: KDI Press, 1974, (in Korean), p. 200. This study's estimates are calculated from Appendix B.

Note: [a] There are important reasons for the differences observed. For the years through 1953 Ban's estimates are more comprehensive, but for the years after 1953, this study's figures reflect a more accurate correction for 1964 changes in crop yield sampling techniques. Ban's estimates to 1953 are to be preferred, since they include the value of livestock production. For other differences see the discussion in Appendix B.

part to different corrections for deficient crop sampling techniques before 1964.

In terms of overall trends in Korea's twentieth-century agriculture, it is also useful to mention that although the farm sector's share of total population has fallen, the farm sector's actual population increased until the mid-1960's, when it finally did begin an absolute decline. In other words, it was not until the beginning of the period of most rapid industrial growth in Korea's economic history that the declining importance of agriculture finally registered as a shrinking in the actual number of people farming. What have been the other changes associated with this decline and ultimate shrinking? In addition to changing crop patterns, there have also been significant farm price movements and shifts in the level and distribution of farm product and income; this is true both between and within regions. An understanding of the most basic aspects of these various topics will greatly expand our general knowledge about Korean agriculture.

Just as the twentieth century has brought structural changes in Korea's economy, greatly reducing the relative significance of agriculture, so has it also brought structural changes within agriculture which have greatly reduced the importance of rice. Much of this change occurred in the years following 1961, though trends in the Japanese colonial period gave indications of similar patterns in certain individual regions of the country. The shift away from rice has in fact resulted from increases in the output of vegetables and livestock products, and these in turn reflect a growing urban population and a growing urban food demand. Table III-3, "Shares of Major Crops, 1918-1975," provides statistics on the actual pattern of changing crop shares for all of South Korea and shows clearly the acceleration in change after 1961. The table provides two sets of data, one calculated using current price data, the other calculated using fixed 1970 prices. Because changes in relative crop prices can be dramatic, particularly in times of bad harvest, the differences between estimates is significant for some years. This significance in relative crop price changes for Korea's rural economy has several principal manifestations, and the following paragraphs will provide a summary account of changes in Korean farm prices in the twentieth century.

Changes in prices that farmers receive for their crops are significant for their relationship to changes in the prices farmers have to pay for goods they use and consume. If prices received move up faster than prices paid, real income from a given harvest is greater. The importance of variations in crop prices is also directly related to the mixture of crops on which any given farmer is most dependent. A sudden fall in the price of a crop to which a farmer has devoted most of his land will be much more harmful than a similar fall in the price of one of his minor sideline products. For these reasons it is important to look at the change over time in the relationship between prices received and prices paid, not only for all of agricultural output but also for major crop categories. A knowledge of these individual terms of trade will be valuable for understanding the differing regional impact of crop price changes.

Tables III-4, "Rice and Barley Purchasing Power Indexes, 1912-1939," and III-5, "Purchasing Power Ratios by Major Commodity Groups, 1959-1975," provide this information for both the Japanese

Table III-3. Shares of Major Crops, 1918-1975

(%)

	Rice	Barley	Pulses	Vegetables	Potatoes	Livestock
1918-20	55.5	12.5	6.7	5.3	1.8	4.3
1928-32	50.3	13.4	5.4	6.9	2.2	4.4
1938-42	51.7 (56.6)	16.2 (13.7)	3.1 (4.6)	5.2 (12.4)	2.0 (1.2)	5.2 (—)
1961-63	54.5 (52.7)	15.4 (10.8)	1.7 (2.8)	5.0 (11.5)	6.0 (3.8)	7.8 (12.0)
1966-68	41.9 (40.3)	14.8 (15.5)	2.4 (2.8)	10.0 (15.9)	5.8 (6.0)	11.5 (10.9)
1973-75	— (41.7)	— (10.3)	— (3.3)	— (16.9)	— (4.1)	— (11.4)

Sources: 1918-1968 data (current prices) are from Sung-hwan Ban, *Hangug Nongeop ui Seongchang (1918-1971)* [Growth of Korean Agriculture], Seoul: KDI Press, 1974, (in Korean), p. 195. 1938-1975 data (1970 prices) from Appendix B.
 Note: The 1918-1968 figures were derived from output figures calculated in current prices. The 1938-1975 data (in parentheses) were calculated from output figures using constant 1970 prices. The 1970 prices exhibit vegetable prices higher than average for the 1960's and 1970's. This difference helps explain the considerable differences in vegetable share estimates.

Table III-4. Rice and Barley Purchasing Power Indexes, 1912-1939

(1934-36 = 100)

Year	Rice	Barley	Year	Rice	Barley
1912	104.1	92.0	1926	101.7	73.3
1913	113.9	77.7	1927	100.3	73.2
1914	93.7	59.4	1928	90.4	82.9
1915	77.7	63.0	1929	91.6	90.5
1916	81.7	61.7	1930	88.6	72.3
1917	94.2	65.0	1931	72.5	42.4
1918	110.4	141.2	1932	85.1	51.0
1919	113.3	66.6	1933	88.1	76.3
1920	103.9	37.3	1934	92.4	80.6
1921	83.4	30.2	1935	104.3	111.0
1922	91.2	28.1	1936	104.2	108.4
1923	86.8	46.3	1937	97.3	99.9
1924	99.2	84.5	1938	92.8	125.0
1925	103.2	96.4	1939	92.1	115.6

Sources: The price of rice is from Bank of Korea statistics reported in Pal-yong Moon, *Nongsanmul Kakyeok Bunseok Non* [Analysis of Farm Product Prices], Seoul: KDI Press, 1975, (in Korean), p. 136. The consumer price index is from Toshiyuki Mizoguchi, *Taiwan Chosen no Keizai Seicho* [The Economic Growth of Taiwan and Chosun], Tokyo: Iwanami Shoten, 1975, (in Japanese), p. 12. The weights used to calculate the consumer index are derived from 1930 Chosun Government General surveys of farm household economies. For a discussion of these and alternative weights see Mizoguchi, cited above, pp. 9-10.

Note: The indexes represent the ratio of prices received by farmers for rice to the prevailing cost of living.

colonial period and the period of rapid industrial growth in South Korea following 1961.

The rice and barley indexes for the colonial period have the drawback of being set in relation to an overall consumer index for Korea, rather than an index of farm village consumption prices only. The data nevertheless show clearly the impact of the world depression on Korean farm terms of trade for the two most important crops. Between 1930 and 1931 the terms of trade for rice deteriorated to levels lower than at any other time covered by the statistics, while barley prices relative to consumer costs fell to their lowest level since the terrible market years of 1920-22. The damage done by these plummeting prices can be imagined more easily when it is remembered that

Table III-5.   Purchasing Power Ratios by Major Commodity Groups, 1959-1975

(1970 = 100)

Year	Rice	Barley	Pulses	Pota-toes	Vege-tables	All Crops
1959	72.5	81.6	64.4	100.6	59.8	72.3
1960	80.7	97.2	75.5	107.0	70.7	81.1
1961	92.7	121.1	69.2	107.5	49.6	88.4
1962	88.8	122.4	66.4	113.1	56.3	90.9
1963	127.5	189.2	99.1	162.5	103.2	120.2
1964	125.0	194.9	119.2	180.4	77.1	126.7
1965	101.4	123.8	109.0	121.0	84.7	104.5
1966	95.4	104.5	105.8	115.7	92.0	98.4
1967	92.8	109.2	123.4	118.3	70.4	96.3
1968	91.2	98.7	77.4	105.0	57.3	88.8
1969	102.5	104.8	74.2	101.2	64.8	94.6
1970	100.0	100.0	100.0	100.0	100.0	100.0
1971	109.8	120.0	91.2	114.2	81.7	104.7
1972	122.2	136.6	97.8	125.8	73.0	111.8
1973	116.7	128.9	104.3	145.7	67.5	110.8
1974	125.4	121.7	97.5	164.9	70.3	112.6
1975	128.5	137.4	98.2	143.0	78.2	114.6

Sources: The indexes were obtained by taking the ratio of (a) output valued in current prices and deflated by a farm consumer price index to (b) output valued in constant 1970 prices. For a detailed discussion and an interpretation of the "weights" implicitly used, see Appendix C.

most of the farm households at that time were tenant households and were in debt. The failing power of a household's ability to repay a fixed obligation meant a rapid drain on a farmer's net proceeds, and it is not surprising that the early 1930's saw both rapidly increasing outmigration from Korean farm areas to Japan and Manchuria and genuine official Japanese concern about the level of "spring famine" and the seriousness of rural pauperism.

Purchasing power data for the 1960's and 1970's are of higher quality because crop prices can be directly compared to the actual cost of living in farm communities. Table III-5 shows that terms of trade generally improved over the course of the seventeen years covered, and it is clear from the changes in the purchasing power of individual crop categories that the prices of rice, barley and potatoes were responsible for the overall improvement, rice and barley most significantly so, given their prominence in overall output.

The changes in prices during the 1960's and 1970's reported in Table III-5 reflect two kinds of influence. The high terms of trade for rice and barley in the 1963-65 period and the extremely high vegetable prices in 1970 result from poor harvests when the supply of output fell far short of demand. The gradual secular increase in the relative prices of rice and barley after 1969, however, was caused by a shift in demand due to growing government support of the prices of both rice and barley. It is significant that there was no such secular movement in the prices for vegetables, the most important crop category after the two principal grains, and this difference in price movements will require further comment in a later chapter, when the importance of price changes for provinces with different crop specializations is analyzed.

Korean agriculture, then, is labor-intensive, rice dependent and yet rapidly changing under the influence of an industrializing economy and an urbanizing society. What has been the effect of all of this on the overall effective productivity of an average Korean farmer? Changes in mean produce per household are of course no sure indication of actual farmer welfare, given the extremes in distribution of land and harvest, but they nevertheless represent the ultimate confluence of changes in prices, yields, farm population and cropping mix. The remaining passages in this chapter will present the uneven pattern of per-household farm output and, in addition, will provide some idea of the degree to which distributional factors qualify its significance.

To what degree did increases in Korean farm output keep up with increases in farm population during the twentieth century? Table III-6, "Per-household Farm Product (1910-1975), 1970 Prices," shows that there were very large increases in the average farm's output at different times during the 65 years documented. Some of the apparent changes, however, are the result of changes in Japanese methods of statistical measurement. The more notable of these was the shift in 1918 to reporting by special officials commissioned from Seoul rather than reliance on local measurement. More accurate reporting after 1917 was also helped by information from the cadastral survey completed in 1918.[3] The other change in sampling techniques occurred

---

[3] For a detailed account of these corrections see Kuro Komikawa, *Chosen Nogyo*

Table III-6. Per-household Farm Product (1910-1975), 1970 Prices

(thousand 1970 won per farm household)

Year		Year		Year	
1910	65	1932	140	1954	192
1911	74	1933	144	1955	200
1912	78	1934	133	1956	181
1913	80	1935	147	1957	199
1914	90	1936	145	1958	213
1915	90	1937	189	1959	211
1916	95	1938	172	1960	198
1917	97	1939	124	1961	233
1918	128	1940	154	1962	206
1919	104	1941	160	1963	207
1920	131	1942	134	1964	252
1921	125	1943	147	1965	241
1922	124	1944	137	1966	273
1923	123	1945	140	1967	256
1924	112	1946	145	1968	260
1925	124	1947	156	1969	295
1926	125	1948	170	1970	288
1927	134	1949	165	1971	299
1928	113	1950	164	1972	307
1929	115	1951	115	1973	318
1930	137	1952	136	1974	340
1931	117	1953	175	1975	358

Source: Calculated from individual crop output and price data and farm household time series. For details see Appendixes A and B.

Note: The 1910-1940 data are for all of Korea; 1941-75 are for South Korea only.

in 1936 and had the greatest significance for rice.[4] Because pre-reform crop reporting is usually a question of underreporting, the changes in sampling techniques invariably involve increases in reported output, and it is important to remember the dates of such changes when interpreting long-term trends in output and per-household output. A third correction in technique occurred in 1964, but the data presented in

---

Hattatsu-shi [History of Chosun Agriculture], Development Volume, Seoul: 1944, (in Japanese), pp. 137ff.

[4] See Choji Hishimoto, *Chosen Mai no Kenkyu* [Chosun Rice Research], Tokyo: Chikura Shobo 1938, (in Japanese), 674-675.

Table III-6 have been adjusted to account for the differences in pre- and post-correction estimates.[5]

A look at the per-farm production levels in Table III-6 shows, however, that it is easy to identify trends in average output which are independent of the changes in sampling techniques. Ignoring figures before 1918, the 1920's were a period of relative stability, when the increases in output were roughly matched by the increase in the size of the farm community. From the year 1930 on, however, production growth was greater than the net increase in farm community size. This increase in average output appeared five years before the change in sampling method, and so is an undisputed shift in relative productivity. We will see in later chapters that this increase had particularly significant regional characteristics.

During the years from the latter 1930's to the early 1950's, through two wars and the political chaos of Japanese repatriation and Korea's division, per-household output roughly held its own, though the highs of the bumper years 1937-38 were never attained until after the Korean War was over. After the Korean War, however, the average production of the South Korean farm began a trend of sustained increase, and as the size of the farm community began an absolute decline in the 1960's, the increases in per-household product greatly accelerated.

We have seen above that changes in the terms of trade of major crops must be important qualifiers for constant-price results such as those just documented, and this is most true for the change in productivity for the early 1930's, when the rapidly falling purchasing power of rice and barley more than compensated for the increases in output. For the post-Korean War period it is possible to see the impact of price changes directly, and Table III-7, "Per-household Farm Product (1959-75), Current Prices," shows that the purchasing-power trends of the 1960's and 1970's, in particular, the increased profitability of rice and barley yielded tangible gross income per farm greater than indicated by the constant 1970 prices. In sum, then, the twentieth century has seen significant increases in actual per-farm output, most notably in the two periods of the early 1930's and the 1960's and '70's. The movements of agricultural prices during these two periods, however, have resulted in different effective impacts for farm income. Average income in the 1930's must have in fact declined

---

[5] For a detailed discussion, see Appendix B.

Table III-7.   Per-household Farm Product (1959-1975), Current Prices

(thousand current won per household deflated
to 1970 by a farm consumer price index)

Year	M.O.A. Survey	(Current Prices) Gross Farm Value Output	(1970 Constant Prices) Gross Farm Value Output
1959	—	152	211
1960	—	160	198
1961	—	205	232
1962	—	187	205
1963	286	249	207
1964	286	319	252
1965	224	252	241
1966	226	269	273
1967	229	246	256
1968	225	230	259
1969	247	279	295
1970	248	288	288
1971	312	313	299
1972	328	342	306
1973	336	352	317
1974	345	382	339
1975	375	410	358

Sources: M.O.A. Survey data are from Ministry of Agriculture and Fisheries, *Farm Household Economy Survey Report, 1975,* 1976. Gross Farm Value Output data were calculated from individual prices and harvest levels for each year. For details see Appendix B. Farm Household data are taken from administrative records summarized in M.O.A., *Yearbook of Agriculture and Forestry Statistics,* various years. Prices and consumer price index data are from N.A.C.P., cited in footnote to Table II-4. Constant price results are from Table II-6.

in spite of the increases in real output, while the increases in product in the 1960's and '70's were amplified by contemporary improvements in terms of trade for major crops.

More important than prices, however, for gauging change in farm community income is the distribution of that income, both between different regions of the country, and more importantly, between individual farms within those regions. If there exists very great dispersion in income, data on overall averages have almost no significance, save perhaps as an indication of the potential for representative farm income. In South Korea's twentieth century development there have been very high levels of inequality, the bulk of it representing differences in the income of individual farms within farm communities, and a much smaller though still significant portion of it resulting from differences between communities, that is, between regions. Some of the available information on these different forms of income inequality is marshalled briefly below.

Differences in average product between large regions of Korea are not perhaps great in comparison to the inequality within the provinces, but the trends in provincial inequality are nevertheless instructive and will be the topic of detailed analysis in subsequent chapters. In general, average farm products of South Korea's different provinces were more equal at the beginning of the century than had become the case by the 1970's. The change, however, was not a smooth one. Table III-8, "Inequality in Per-household Farm Product between Provinces, 1921-1975," shows that inequality between provinces increased very rapidly from the near equality of the twenties to the years of greatest inequality leading up to and during the War in the Pacific. Thereafter the extremes in differences between the provinces quickly declined, but the equality of the early decades in the century were never again attained. Although the post-Korean War years seem rather uninteresting when viewed in this summary way, a later detailed look at the changes in relative status of individual provinces will show important phenomena hidden beneath these figures.

But what of the major source of inequality in rural Korea? What measures have we of differences in the situations of individual farms? The greatest differences between poor and rich farmers clearly were those of the Japanese colonial period. The Japanese kept no statistics on farm income by size, but the data in Table III-9, "Farm Income Distribution by Farm Type and Size, 1925," provide very similar information and, in addition, differentiate the farms by their tenant or landlord status. The extreme poverty of tenants and small tenant owners is remarkable, and these farmers represented over 60 percent of the total. The landlords, however, made up only 4.5 percent of the population, and yet received roughly 13 percent of all gross product and a full 52 percent of all farm net income.[6] It is likely, however,

---

[6] Calculated from population figures and average income data in Table III-9.

*Table III-8.   Inequality in Per-household Farm Product between Provinces,
1921-1975*

(Theil's measure[a])

Years	Index
1921-25	.004
1926-30	.004
1931-35	.012
1936-40	.031
1941-45	.037
1946-50	.011
1951-55	.012
1956-60	.009
1961-65	.009
1966-70	.008
1971-75	.008

Sources: Calculated from data on average provincial household product which were in turn derived from provincial crop statistics and individual crop prices. For details see Appendixes A and B.

Note: [a] Theil's measure is derived from information theory and summarizes the degree to which given shares of the population receive commensurate shares of total product or income. For a detailed discussion see Henri Theil, *Economics and Information Theory*, Chicago: Rand McNally, 1967, pp. 91-106. An index of zero implies perfect equality; increasingly large indexes imply increasing inequality. The measure is not bounded from above and changes with the size of the sample, but it has the important advantage of ease of decomposition (see Table III-12 and its accompanying explanation).

that these data overstate the poverty of the tenants and smaller tenant owners. The average receipts for all farms is 66.7 thousand 1970 won, which is well below the average per-farm value of produce for the same year (124 thousand 1970 won; see Table III-6), even when allowances are made for changes in relative prices (the results from Table III-9 are current 1925 prices deflated to 1970 equivalents) and the possibility of bias in the sampling technique. This difference indicates that much of farm income in kind, that is, produce not sold but rather consumed on the farm, was not included in the Japanese survey used as a basis for Table III-9. Because small farms generally consume directly a larger portion of total produce than is the case for larger and more commercial farms, the results from a survey of cash receipts and incomes will underestimate the actual welfare of poorer farmers. Nevertheless, the figures in Table III-9 are dramatic and

Table III-9. Farm Income Distribution by Farm Type and Size, 1925

(households; thousand 1970 won per household)

	Households	Share of Total Households	Average Receipts	Share of Total Receipts	Average Costs	Average Income
Landlords						
Large	6,886	.3	1,401.9	5.3	671.4	730.6
Medium	22,994	.8	292.6	3.7	200.5	92.1
Small	39,455	1.4	124.9	2.7	93.4	31.4
Tiny	52,670	1.9	61.1	1.8	55.0	6.2
Average*/total	(121,985)	(4.5)	(200.9)	(13.5)	(129.5)	(71.3)
Owner-tenants						
Large	94,453	3.5	162.0	8.4	131.4	30.5
Medium	179,016	6.6	95.8	9.4	83.1	12.7
Small	172,390	6.3	57.7	5.5	52.5	5.2
Tiny	107,819	4.0	41.1	2.4	38.9	2.2
Average*/total	(553,678)	(20.3)	(84.6)	(25.7)	(73.2)	(11.4)
Tenant-owners						
Large	98,628	3.6	132.8	7.2	120.9	11.9
Medium	263,747	9.7	77.9	11.3	72.1	5.8
Small	329,431	12.1	49.9	9.0	48.9	.9
Tiny	225,605	8.3	31.5	3.9	31.7	-.1
Average*/total	(917,311)	(33.6)	(62.3)	(31.4)	(59.1)	(3.2)

Table III-9.   (Cont'd)

	Households	Share of Total Households	Average Receipts	Share of Total Receipts	Average Costs	Average Income
Tenant						
Large	88,226	3.2	107.8	5.2	105.8	2.1
Medium	233,029	8.5	77.3	9.9	78.0	−.7
Small	354,399	13.0	43.6	8.5	46.2	−2.6
Tiny	298,084	10.9	28.1	4.6	29.7	−1.6
Average*/total	(973,738)	(35.7)	(52.8)	(28.3)	(54.2)	(−1.4)
Pauper	162,209	5.9	13.4	1.2	13.9	−.5
Overall Average*/total	2,728,921	100.0	66.7	100.0	60.7	6.1

Source: Chosun Government General, *Chosen no Shosaku Kanshu* [The Practice of Tenancy in Chosun], Seoul, 1928 (?), (in Japanese), pp. 33, 38. The conversion to 1970 won from 1925 yen was made on the basis of the three-year average prices of rice, centered around 1925 and 1970. The conversion factor was thus 130.88 1970 won/1925 yen.

Note: The size designation in these categories refers both to land owned and to land managed. The receipts and income figures refer to the economies of households rather than of identifiable tracts of land. A landlord owning and receiving rents from much land is a large landlord, and a tenant owning no land but managing much land is a large tenant.

* The average used is a weighted one, with the share of farm households in each category used as weights.

reveal the desperate living conditions of the majority of Korea's colonial farm community.

A contrasting picture of dramatic change for the better is provided by Table III-10, "Farm Income Distribution by Size of Holding, 1975," which also provides information on the pre- and post-land-reform distribution of land by size. As can be seen from the first six columns of this table, the share of farm households having farms over two hectares in size, fell from 10 percent to 6 percent between 1945 and 1975, and more significantly, the share of land held by these larger farms decreased from 62 percent of all land to only 21 percent. The land reform, then, radically distributed land away from the larger holdings to the great benefit of the intermediate-sized farming class. It is interesting that neither the share of population nor the share of land in the lowest category was much affected, as shown by the stability of both shares indicated in Table III-10.

This same table shows the different gross and net incomes of farms in these several size categories. The larger farms do better, of course, but the extremes are nowhere near what they were for the colonial period, though the 1975 data are much less detailed. In sum, this brief view of farm income by farm size for Korea in the 1970's shows great improvements over the 1920's, and the role of the 1949 land reform seems to have been crucial in this change, not only for the change shown here in actual size of holdings, but also because the land reform meant a great reduction in the widespread tenancy and tenancy obligations of the Japanese period not revealed in land ownership statistics.

The singular importance of the land reform is emphasized when it is shown that the pattern of land ownership inequality has changed very little since the end of the Korean War. This point is made clear by Table III-11, "Between and Within Province Inequality in Land Ownership, 1955-75," which also uses a valuable feature of the measure of inequality chosen, Theil's measure. This table shows that compared to 1945-46, when the land inequality indexes were .854 and .925 respectively, inequality after 1955 remained between .225 and .267. The difference between 1945 and 1975, then, is a result of the land reform only, and inequality since 1955 has in fact tended to increase again slightly.

The valuable feature of Theil's measure of inequality is its easy decomposition into portions due to differences between regions or groups on the one hand, and differences among households within

Table III-10.　Farm Land and Income Distribution by Size of Holding, 1945 and 1975

(thousand households; thousand 1970 won)

Farm Size (hectare)	1945 No. of Households	1945 Share of Total Households	Land Held as Share of Total	1975 No. of Households	Share of Total Households	Land Held as Share of Total	1975 Agricultural Gross Receipts	1975 Agricultural Expenditures	1975 Net Receipts
2.0 <	202	(10.0)	(62.3)	148	(6.2)	(20.5)	1,086.3	254.3	832.0
1.0 – 2.0	459	(22.9)	(14.1)	618	(26.0)	(41.2)	645.5	130.0	515.6
0.5 – 1.0	671	(33.4)	(12.3)	828	(34.8)	(28.3)	392.9	75.3	317.6
< 0.5	677	(33.7)	(11.3)	785	(33.0)	(10.0)	194.1	32.9	161.2
Total/Ave.	2,009	(100.0)	(100.0)	2,379	(100.0)	(100.0)	375.0	74.1	300.9

Sources: The 1945 data are from Bank of Chosun, *Chosun Kyeongje Yeonbo* [Chosun Economic Yearbook], Seoul, 1948, p. I-31. 1975 Household data are from Ministry of Agriculture and Fisheries, *Yearbook of Agriculture and Forestry Statistics, 1975* and *Farm Household Economy Survey Report, 1975* both published in Seoul, 1976. The average is weighted by shares of households in land holding categories, except for income averages, which are from the household survey's results. The 1975 values were deflated to 1970 by a farm consumer price index. See the note on Table II-4.

Note: The 1945 land-holding shares were estimated by taking the mid-points of land-holding size categories multiplied by the number of households in the category. Because the mid-points differed from the (unknown) averages, the resulting shares were normalized to add to 100 percent.

Table III-11. *Between and Within Province Inequality in Land Ownership,*
1955-1975

(Theil's measure[a]; percent)

Year	Inequality	Between Provinces[b]		Within Provinces[b]	
		Inequality	(share)	Inequality	(share)
1945	.854	—		—	
1946	.925	—		—	
1955	.225	.014	(6.2)	.212	(93.8)
1960	.248	.008	(3.4)	.239	(96.6)
1965	.243	.007	(3.0)	.236	(97.0)
1970	.267	.008	(2.9)	.259	(97.1)
1975	.261	.007	(2.8)	.254	(97.2)

Sources: Land distribution data for 1945-46 are from Bank of Chosun, *Chosun Kyeongje Yeonbo* [Chosun Economic Yearbook], Seoul, 1948, pp. I-32 to I-36. For 1955-75 the data are from provincial administrative land registers as reported in individual year issues of the Korean Ministry of Agriculture and Forestry, *Yearbook of Agriculture and Forestry Statistics,* Seoul.

Notes:[a] See footnote to Table III-8.

[b] For an explanation of these categories, see the text.

those regions or groups on the other. More specifically, Theil's measure of total inequality in an economy (TM_T) is the sum of Theil's measure of inequality between group averages (TM^*) and a weighted sum of Theil's measure of inequality between individual units within each group (TM_j for group j), where the weights are the provincial shares in total population (s_j).[7] The decomposition of Theil's measure, then, can be summarized in the following expressions:

$$TM_T = TM_{between} + TM_{within}$$
$$= TM^* + \sum_j s_j \, TM_j .$$

Table III-11 shows that the overwhelming proportion of inequality in land holding is due to different farm sizes within each province, rather than between-regional differences in average holdings. Over time, between-province differences have shown some decline, while

[7] For a detailed discussion of Theil's measure and its decomposition see Henri Theil, *Economics and Information Theory,* Chicago: Rand McNally, 1967, pp. 91-106.

Table III-12.   *1970 South Korean Agricultural Inequality Between Provinces, Between Counties and Within Counties*

(Theil's Measure[a])

Category	Total Inequality	Between Provinces[b] Inequality (%)	Between Counties[b] Inequality (%)	Within Counties[b] Inequality (%)
1. Average Farm Size	.286	.010 (3.5)	.008 (2.8)	.268 (93.7)
2. Average Cash Sales	.759	.012 (1.6)	.036 (4.7)	.711 (93.7)
3. Average Vegetable Sales	1.701	.104 (6.1)	.212 (12.5)	1.385 (81.4)

Source: Calculated from data in Republic of Korea, Ministry of Agriculture and Fisheries, *Agricultural Census, 1970*, Seoul, 1974.

Notes:[a] For an explanation of this measure see the note to Table III-8.

[b] Inequality between provinces measures differences between provincial averages; that between counties is calculated from inequality within provinces based on county averages, summing these within-province measures using provincial population weights; within-county inequality is the weighted sum of individual county inequality levels weighted by county shares of farm population.

within-province inequality increased slightly between 1955 and 1975. These movements, however, do not seem significant when compared to the impact of the land reform.

If the bulk of inequality in land holding is within provinces, is that also true for other indexes of farm size such as output and income? If so, a study of regional differences would seem to offer precious little information about the development and changing status of Korea's rural economy. Table III-12, "1970 South Korean Agricultural Inequality," shows that low levels of interregional inequality are much less characteristic of cash sales and vegetable sales than they are of land, and that when regional units smaller than provinces (counties) are used, a larger share of overall inequality can be isolated for examination.

The data in Table III-12 are from an agricultural census rather than from administrative reporting, and yet the overall and between-province measures of inequality are comparable. Because the data are available by county as well as by province, Theil's measure can be used to calculate a three-way decomposition.

It is interesting that differences in land holdings between counties are even less significant (.008) than those between provinces (.010), but that this pattern does not hold for inequality either in total cash sales or in vegetable sales. For cash sales, inequality levels between provinces and between counties are both greater relative to land, but only between-county inequality increases its share of total inequality in relation to land holding, since inequality in cash sales overall is itself considerably greater.

Even greater inequality is apparent in the sale of vegetables, and it is significant that a considerable percentage of the total inequality is due not only to between-county differences (12.5 percent) but also to between-province gaps (6.1 percent). This is particularly significant when it is remembered that increased cultivation of vegetables is the single most significant mark of change in the output mix of Korean agriculture, and it is hence clear that a study of these changes at the county level, and even at the province level, could provide a great deal of insight into the actual working of Korean agriculture, especially after the Korean War.

Finally, supplementary information about land, cash sales and vegetable sales is provided by the degrees of correlation between county averages for these variables and measures for the inequality in their distribution within each county. Table III-13 provides a matrix of simple correlation coefficients. These figures are most interesting for showing the impact of vegetable cultivation on overall county inequality. As might be expected, inequality in cash sales correlates positively with inequality in farm size (.672), but it is interesting that the same relationship between vegetable sales and size of holding does not exist (.058). This is more easily understood when it is noticed that the table's most significant coefficient (−.698) is between the level of vegetable sales and the degree of inequality in household receipts from vegetable sales. In other words, where there are more vegetable sales, there is also greater equality in the distribution of income from vegetable sales. This implies that vegetable cultivation has a leveling effect on the distribution of farm income, and it is intriguing that this

Table III-13.   Correlation of 1970 County Averages and Inequality Levels

Symbol	Meaning
L/HH	Average household landholding (farm size)
Cash/HH	Average household cash sales of all Agricultural Products
Veg/HH	Average household cash sales of vegetables
Q-L	Within-county inequality (Theil's Measure) in land holding
Q-Cash	Within-county inequality in cash sales
Q-Veg.	Within-county inequality in vegetable sales

	L/HH	Cash/HH	Veg/HH	Q-L	Q-Cash	Q-Veg.
L/HH	1.000	.221	.044	− .460	− .402	− .141
Cash/HH	.221	1.000	.547	.331	.139	− .355
Veg/HH	.044	.547	1.000	.244	.160	− .698
Q-L	− .460	.331	.244	1.000	.672	.058
Q-Cash	− .402	.139	.160	.672	1.000	.117
Q-Veg	− .141	− .355	− .698	.058	.117	1.000

Sources: The averages and the Theil inequality measures were calculated from data in ROK, Ministry of Agriculture and Fisheries, *Agricultural Census, 1970,* Seoul, 1974.

single most important factor influencing changes towards more inequality between regions in the post-Korean War era is at the same time a factor encouraging less inequality within those regions most benefited. This result is confirmed by noting that there is in effect no correlation (.044) between the average size of farms and the average level of vegetable sales. The distribution, then, of farm land and farm produce has changed profoundly during the course of the twentieth century, and has been influenced by forces as direct as the land reform and as complex as the spread of vegetable cultivation. Although it is true that the bulk of the inequality exists within individual regions, there is also a large enough portion reflected in differences between regions to ensure fruitful results from their detailed examination.

The state of South Korea's agriculture as outlined in this chapter has been one of flux. Changes in total output, cropping mix, per-household income and the distribution of that income have all indicated an evolution in the direction of an agricultural sector more suited to the needs of a modernizing society. Korea's rural economy is unique among developing economies for the density of its farm population and the intensity of its rice cultivation. Nevertheless, the twentieth-century pattern of its agricultural growth can provide important lessons for rural development in other advancing economies. Studying regional trends within a single economy provides such an increase in detail and variation when compared to an examination of national total statistics alone, that the remainder of this study will turn to the individual experiences of South Korea's eight mainland provinces and two special cities (Seoul and Pusan). As we shall see, some of these regions have been in the vanguard of changes later characteristic of the overall economy, while other regions have lagged far behind.

CHAPTER IV

INTRODUCTION TO THE REGIONS OF SOUTH KOREA

Although small and by many measures quite homogeneous, South Korea has several distinct regions whose characteristics result from significant differences in climate and terrain. Because these regions will be used as the basis for most of the remaining analysis in this study, they deserve careful attention. In addition to reviewing the important influences of mountains, temperature and rainfall, the present chapter will also detail the regional differences in agricultural land and the location of the principal urban areas in relation to the pattern of natural endowments.

When discussing the regions of South Korea and their individual characteristics, however, it is important to remember that South Korea is a geographically small country, and that compared to other developing economies it enjoys a considerable degree of homogeneity between its various sections. The distance between its two principal cities, Seoul and Pusan, the one in the extreme Northwest and the other in the extreme Southeast, is 210 miles (340 kilometers) or less than the distance between New York and Washington. In addition, the homogeneity of rural Korea is in stark contrast to conditions in most other developing countries, such as Brazil, Indonesia, Malaysia and India. In those economies, racial, linguistic and cultural differences between regions inject external factors hindering economic analysis, and extreme crop specialization in different areas accounts for much of the variety in regional production and income. These extremes in regional variation are absent in South Korea.

Of almost complete racial homogeneity, Koreans everywhere speak essentially the same dialect of their Ural-Altaic-related language with only some regional differences in pronunciation (for example, the natives of South Chungchong Province are well known for their slow delivery, and yet listening to them speak one is aware of differences less marked than those of a Mississippi drawl as heard by a New England Yankee). In addition to race and language, the regions of South Korea share a common farm technology and plant the same basic mix of crops throughout the peninsula. As we have seen, rice is the much preferred foodgrain, and it is traditionally planted in all parts of the country where land and irrigation are suitable. Furthermore, a common set of secondary crops (barley, pulses, potatoes, cabbage, cucumber and radish) is extensively planted in all provinces. This fundamental homogeneity in race, language and technology, then, in addition to being one of the sources of potential strength for the Korean economy, also provides a valuable base for interpreting those differences between regions which have appeared. In Korea's case, regional variation is much more likely to be the result of economic forces than of outside random phenomena, and economic analysis of regional differences and trends has a good chance of uncovering important motors of change in the country's rural economy.

Natural conditions in South Korea are best viewed in the context of two composite influences: mountains and weather. In general, the East of Korea is rugged, with mountains which begin at the Sea of Japan and render most land untillable. The mountains and hills become less severe the further west one goes, until finally one reaches the plains and alluvial areas facing the Yellow Sea. Superimposed on this East-West scale is the predictable South-North temperature range, which allows natural winter cropping in southern provinces, but not in northern ones. The volume and timing of rainfall are also most propitious in the South and become less so in more northerly areas.

The distribution and severity of mountains can be seen in Map IV-1, "Mountains of South Korea." The southern part of the Taebaek Mountains extends the length of the peninsula, and although its eastern portion becomes less severe in the more southerly regions, a splinter range juts out to the Southwest and penetrates almost as far as the sea. It would be difficult to exaggerate the significance of these mountains for Korea's rural economy. In addition to directly influencing the availability of arable land, mountains form barriers to

Map IV-1.   Mountains of South Korea
(Higher numbers represent a higher proportion of total county area in
mountains.)

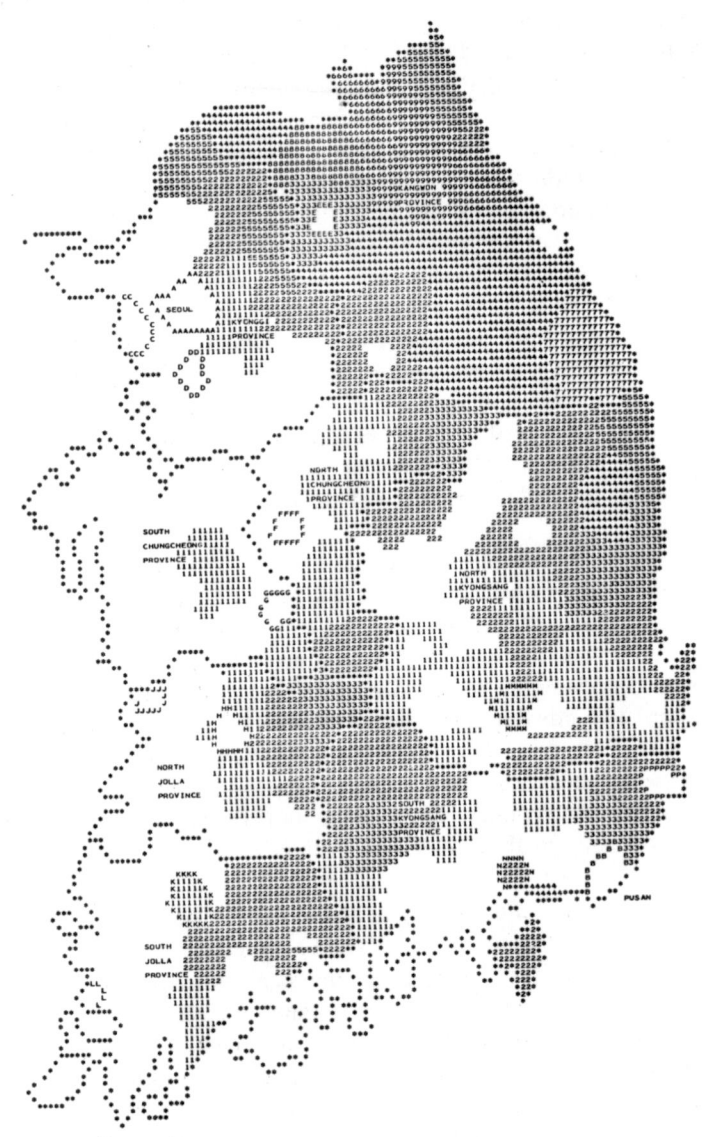

Map IV-2. Mean Annual Air Temperature (°C)

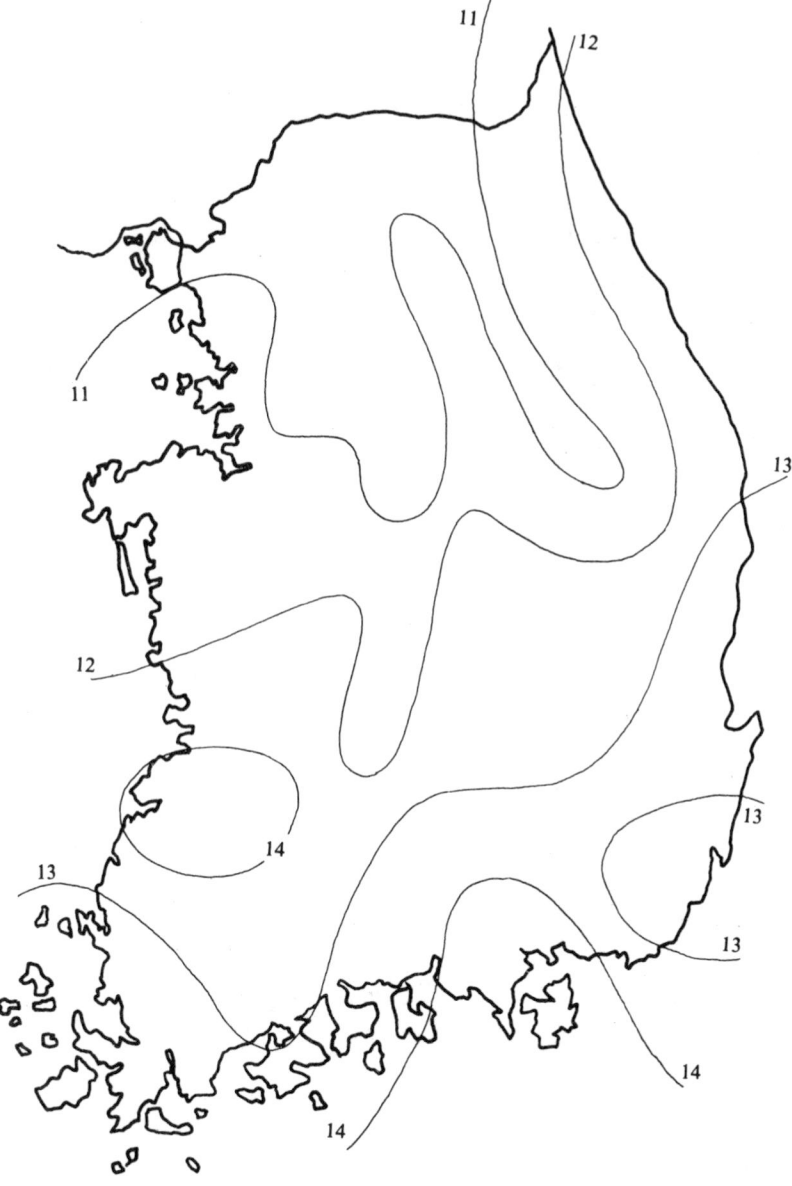

Map IV-3.   Last Date, Frost

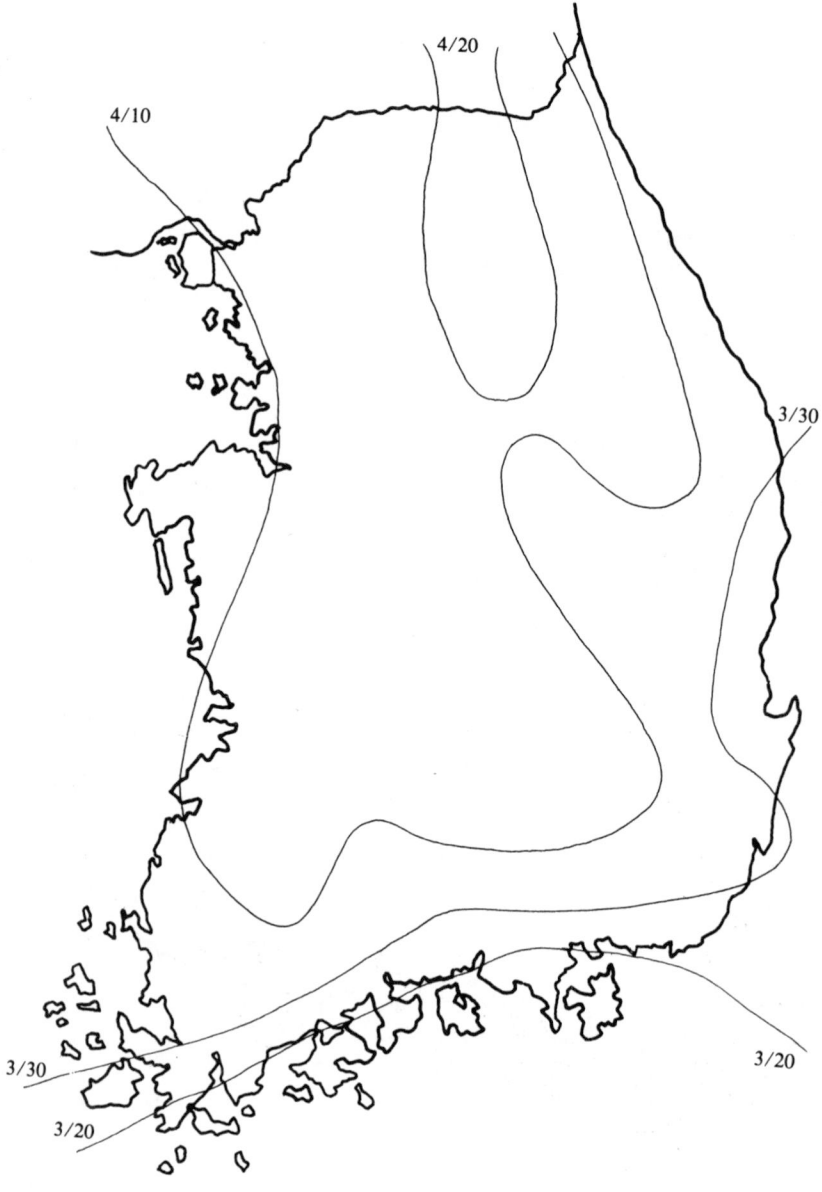

communication and transport and exacerbate the cold and early frosts common for Korea's latitude.

The level and regional variation in Korea's temperature can be seen in Maps IV-2 and IV-3, "Mean Annual Air Temperature" and "Last Date, Frost."[1] The South and particular regions in the Southwest have the warmest temperatures overall, and the mean temperature is clearly lower in northern areas. The significance of these seemingly slight differences in annual mean can be understood when it is noted that summer temperatures are in essence equal over most of the country. The differences in annual average, then, represent the difference in times of seasonal change in northern and southern sections and also differences in the severity of winter. Map IV-3 gives some idea of the difference in seasonal change, for there is a month's difference between the last frost of Spring in the South and that in the North. The contours for first dates of Fall frost are quite similar to those of Map IV-3, the earliest frosts in the North coming the tenth of October while in the South not until the tenth of November, again a difference of a month. In sum, as measured by frosts, the growing season in the coldest Northeastern regions is a full two months shorter than in the southernmost parts of the country.

The benefits of early warmth in the South are complemented by South Korea's overall pattern of rainfall. As shown in Map IV-4, "Annual Precipitation," except for a few mountainous areas the South enjoys greater total rainfall, especially along the southernmost coast and in the Southwest. But here, as for temperature, the timing of precipitation is as important as the overall average, and other statistics show that rainfall in the early part of the growing season (March through May) is considerably heavier in the South than in the North, while the difference is made up during the months of heaviest rain (June through August).[2] In other words, both temperature and rainfall favor the South, and when this general pattern is considered in conjunction with the location of mountains, the overall natural endowment of different quadrants of the country becomes easy to understand.

Joint consideration of the severity of mountains and the variation in temperature and rainfall yields five crude zones in South Korea.

---

[1] These and following climatic maps are from Central Meteorological Office, *Climatic Tables of Korea (1931-1960), Climatological Standard Normals, Part I,* Seoul, 1968.

[2] *Ibid.*

Map IV-4.   *Annual Precipitation (millimeters)*

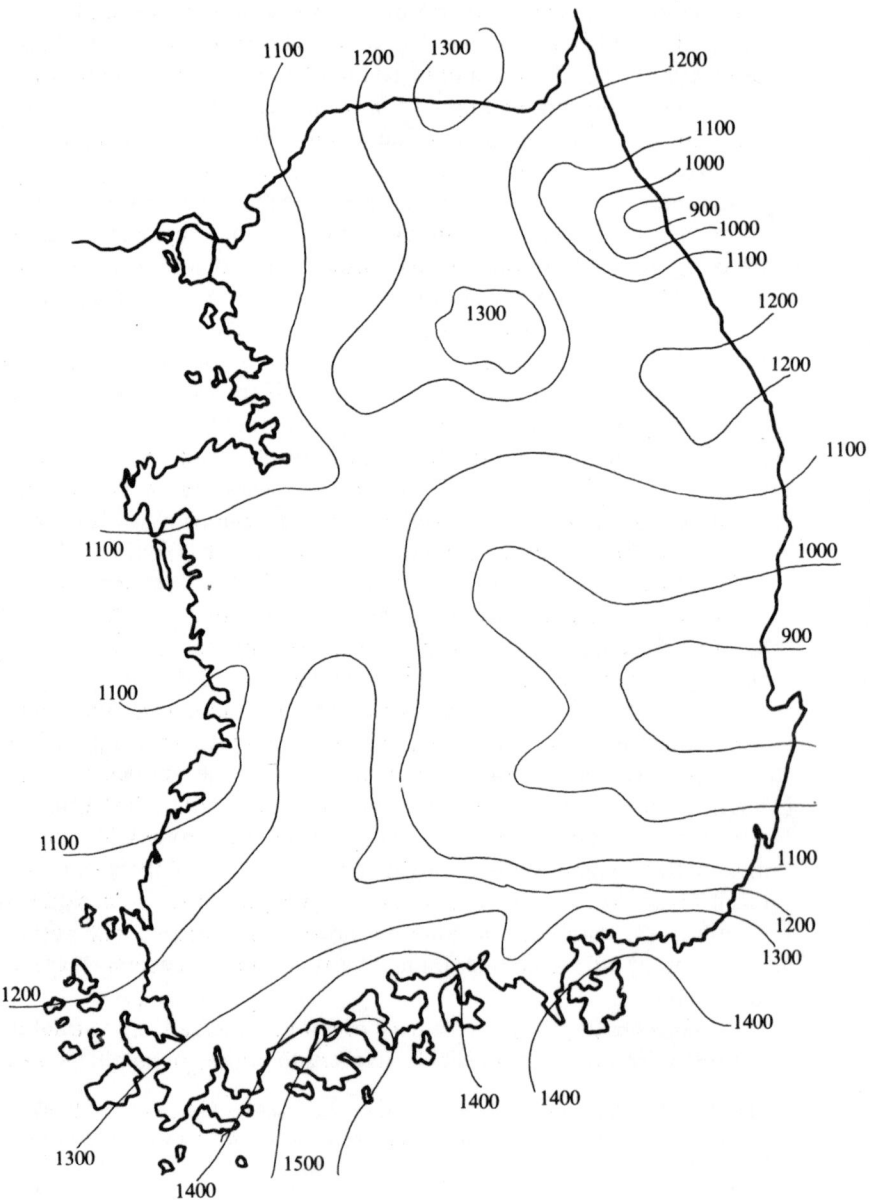

The Southwest, with the warmest temperatures and fewer mountains, has received nature's best combination. The Northeast, on the other hand, has all of the worst attributes: early and cold winters, late springs, and rugged terrain yielding only pockets of arable land. A third region in the Northwest has much good land, but is not so warm, while the Southeast combines mountain valleys with warmth and some of the best rainfall conditions of all. Finally, the central area, where attributes of all the above four regions blend, is endowed with hilly farmable land and moderate temperatures, but less of the paddy land found to the West and shorter growing seasons than found to the South.

Although there are no administrative boundaries which naturally conform to the climatic and geographical regions described above, it is convenient to refer to the provinces of South Korea because they form the basis for regional statistics and because discussion of regional phenomena in South Korea is generally in terms of provinces or groups of provinces. Map IV-5, "Provinces and Cities of South Korea," is provided for reference in the following discussion. The smaller lettered areas are manufacturing cities. (For an explanation of the letter codes see the note to Table IV-5 on page 70.)

The poorest region in South Korea throughout most of the period of this study has been the very large northeastern area made up of Kangwon Province. Known popularly as "Potato Country," Kangwon is extremely mountainous, and as we have seen, combines the worst of the climatic and geographical conditions in South Korea. Transportation and communication within the province and with other parts of the country has traditionally been very difficult. Rice is not easy to grow (hence the importance of alternative staples), and side occupations in forestry and mining help supplement farm income.

Contrasting with Kangwon in the Northeast is the traditional Korean rice belt represented by the three provinces which contain the Southwestern plains and valleys. South Chungchong and North and South Jolla enjoy less frost, early rainfall and generous endowments of arable land, most of which is suitable for wet-paddy rice cultivation. In addition, warm temperatures allow farmers in the southern provinces to double-crop winter barleys on much of their land, thereby greatly increasing the area's total annual yield. We will see, however, that these same southwestern provinces, in particular the two Jollas, have been handicapped by poor transportation to and

Map IV-5.   Provinces and Cities of South Korea

(Letter boundaries around towns and cities are identified in Table IV-5, page 70.)

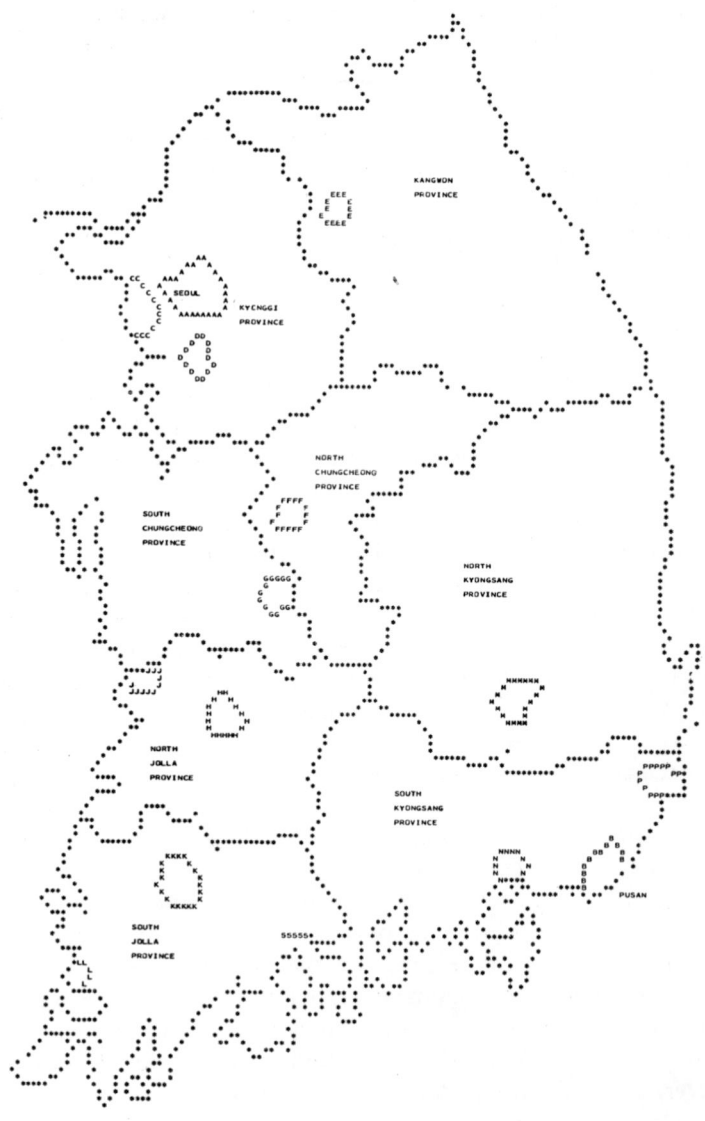

communication with the urban and industrial areas to the north and east.

The regions in the Northwest and Southeast are in one sense hybrid regions between the two extremes of Korea's best and worst described above. The Northwest is represented by Kyonggi Province and by the Special City of Seoul. The wealthiest of South Korea's farmland regions through much of the twentieth century, it has a lower density of farm population per hectare of cultivated land when compared with other areas, and is a hybrid region because while it has taken good advantage of the profitability of vegetables, fruits and livestock, the importance of rice in total output and income came to approach that found in the rice bowl to the south.

South Kyongsang Province and Pusan (together with southern portions of North Kyongsang Province) best represent the hybrid region to the Southeast. Embracing South Korea's second Special City, the major port of Pusan, South Kyongsang is, however, a hybrid quite different from Kyonggi in the Northwest. It has by far the highest density of farm population per cultivated hectare in all of South Korea. Like Kyonggi, South Kyongsang has shown a rapid increase during the twentieth century in the importance of vegetables and also relies heavily on rice production, but because it is more mountainous, it has many sections without access to Pusan or other markets, and considerable shares of its total output are in barleys and potatoes, a characteristic of the most isolated regions.

The remaining central region is best typified by North Chungchong province but includes portions of North Kyongsang as well. Throughout the rest of this study this region will be referred to as the "upland" region because of the predominance of hilly dry fields in comparison to the southern and western provinces, although paddy land still represents over 40 percent of all cultivated land. At a traditional disadvantage because of its relatively little paddy, this upland region will be the focus of attention during the analysis of farm produce and income during South Korea's period of rapid economic growth in the 1960's and 1970's.

The above brief description of South Korea's principal mainland rural regions has left out the island province of Jeju. Jeju Island is well to the south of the Korean Peninsula, and with its very different climate and volcanic origins it shares neither the emphasis on rice cultivation found throughout the mainland nor the severe winters and shorter growing seasons of the other regions. Very poor for much of the century, Jeju began rapid development of its orange industry in the latter 1960's, and in very recent years this single crop has given the island the highest average per capita agricultural output in all of South Korea. Because of these special conditions, Jeju sets itself apart from the more homogeneous mainland agricultural regions and is difficult to include in a meaningful comparison of trends and patterns in rural development. Therefore, although data for Jeju will be presented with other provincial statistics, in subsequent chapters the island's economy will play no role in accompanying analysis.

Having identified Korea's major regions with individual provinces or groups of provinces, it is now possible to further support their description with provincial data. Table IV-1, "Area and Mountains of South Korean Provinces," confirms that the western provinces have fewer mountains than those in the east. Kyonggi and the three rice bowl provinces all have less than 70 percent of their area in mountains, while none of the others save Jeju does. Furthermore, the overwhelmingly large share of Kangwon's land in mountains (89.5 percent) clearly sets that region apart. Table IV-2, "Cultivated Land by Province, 1920-1975," shows that Kyonggi and the rice belt provinces account for roughly 55 percent of all cultivated land, while calculations from Table IV-1 show that these same provinces have only 40 percent of the nation's total land. Table IV-2 is also interesting for revealing the degree to which cultivated land fell out of cultivation during the Second World War. Because of boundary changes, this result is unobtainable from national statistics, yet time series for province after province show it to have clearly been the case. In other words, although there has been an overall increase in cultivated land since the Korean War in South Korea, the effect has been to return the cultivated areas to their level under the Japanese, and in this sense, there has been very little increase in total cultivated land during South Korea's twentieth century.

Finally, Tables IV-3 and IV-4 show the regional endowments in all-important rice paddy and the degree to which that paddy was planted in two crops annually. Table IV-3, "Provincial Paddy as a Share of Cultivated Land, 1920-1975," shows the degree to which the rice belt and hybrid provinces are endowed with paddy. In spite of its mountains, the Southeast province of South Kyongsang is second only to North Jolla and has two-thirds of its farmland in rice. Again, the relatively poor gifts of Kangwon are revealed, for the poor Northeastern region has only 37 percent of its arable land in paddy. Table IV-3 also

Table IV-1.   *Area and Mountains of South Korean Provinces*

Province	Total Area (km²)	Mountains as a Share of Total Area (%)
Kyonggi	11,071	67.0
Kangwon	16,827	89.5
North Chungchong	7,436	73.8
South Chungchong	8,764	61.0
North Jolla	8,058	64.3
South Jolla	12,084	65.4
North Kyongsang	19,805	77.8
South Kyongsang	11,966	74.1
Jeju	1,820	68.2
South Korea	98,807	73.5

Source : Total area and non-cultivated land use: Economic Planning Board, *Korea Statistical Yearbook*, 1976, pp. 17-22.

Table IV-2.   *Cultivated Land by Province, 1920-1975*

(km²)

Province	1920	1935	1949	1955	1965	1975
Kyonggi[a]	3,900	3,950	3,994	2,870	3,162	3,072
Kangwon[a]	3,620	4,340	1,126	1,281	1,570	1,549
N. Chungchong	1,620	1,630	1,387	1,443	1,655	1,764
S. Chungchong	2,450	2,490	2,323	2,469	2,872	2,919
N. Jolla	2,390	2,430	2,261	2,385	2,524	2,503
S. Jolla[b]	4,170	4,290	3,075	3,277	3,774	3,617
N. Kyongsang	3,920	3,860	3,419	3,465	3,792	3,817
S. Kyongsang	2,800	2,780	2,484	2,513	2,735	2,660
Jeju	—	—	366	382	478	495
South Korea[a]	24,830	25,740	20,535	20,084	22,564	22,397

Sources: 1920 and 1935 data are from Chosun Government General, *Chosun Statistical Yearbook*, individual years. 1949-75 data are from Ministry of Agriculture and Forestry, *Yearbook of Agriculture and Forestry Statistics*, relevant years.

Notes: [a] These provinces and totals are not comparable between 1935 and 1949 or between 1949 and 1955 because of changes in provincial boundaries following the division of Korea in 1945 and the Korean War Armistice of 1953.

[b] Jeju Island was part of South Jolla until after World War II.

Table IV-3.   Provincial Paddy as a Share of Cultivated Land, 1920-1975

(%)

Province	1920	1935	1955	1965	1975
Kyonggi[a]	51.5	53.2	61.3	60.5	60.8
Kangwon[a]	21.8	21.4	35.4	35.7	37.3
N. Chungchong	43.8	44.8	48.3	46.4	45.0
S. Chungchong	66.1	66.3	66.7	62.0	61.7
N. Jolla	71.1	71.6	73.9	67.1	68.0
S. Jolla	49.2	49.9	62.7	58.9	59.4
N. Kyongsang	48.0	51.0	56.0	54.9	54.9
S. Kyongsang	57.9	65.1	67.0	66.5	66.7
Jeju	—	—	2.1	2.1	2.1
South Korea[a]	49.9	50.7	59.8	57.0	57.0

Sources: 1920, 1935 data are from Chosun Government General, *Chosun Statistical Yearbook,* selected years. 1955-75 data are from Ministry of Agriculture and Forestry, *Yearbook of Agriculture and Forestry Statistics,* individual years.

Note: [a] Because of changes in boundaries of these two provinces resulting from Korea's division, the data are not comparable between 1935 and 1955.

Table IV-4.   Double-cropped Paddy as a Share of Total Paddy, 1915-1975

(%)

Province	1915	1927	1935	1955	1965	1975
Kyonggi[a]	.6	.6	2.3	3.4	9.1	25.7
Kangwon[a]	1.0	1.2	3.1	2.5	4.2	6.7
N. Chungchong	18.6	21.7	27.1	31.1	35.0	45.6
S. Chungchong	7.3	12.0	17.1	32.2	32.6	56.6
N. Jolla	10.7	21.5	38.6	48.8	50.3	76.3
S. Jolla	18.8	32.3	52.7	50.0	58.8	79.8
N. Kyongsang	42.5	40.2	49.0	45.2	62.3	61.7
S. Kyongsang	32.3	43.2	57.7	53.8	71.1	81.9
Jeju	—	—	—	25.8	40.7	52.4
South Korea[a]	17.0	23.1	33.4	37.4	45.1	60.0

Sources: 1915, 1927, 1935 data are from Chosun Government General, *Chosun Statistical Yearbook,* individual years. 1955-75 data are from Ministry of Agriculture and Forestry, *Yearbook of Agriculture and Forestry Statistics,* individual years.

Note: [a] Kyonggi, Kangwon and South Korea are not comparable between 1935 and 1955 because of boundary changes at the time of Korea's division.

provides valuable time-series information linking the Japanese period with years following the Korean War. In general, the share of land planted to paddy increased until the mid-1950's, but began a slight decline during the following decade, arrested perhaps only by the change in terms of trade for barley and rice, both grown on paddy, of the late 1960's and early 1970's. Paddy is used for both these crops in some areas because although it is too cold in Korea to double-crop rice, it is possible to plant a winter crop of barley after the autumn rice harvest, and this second crop is then harvested late in the following spring in time for the rice transplanting. Table IV-4 shows the degree to which this "double-cropping" has been practiced in South Korea, and it is clear at once that southern provinces have had an advantage in this respect. It is also clear that in the course of the century double-cropping has expanded greatly in all areas. To some degree this represents the extension of traditional double-cropping techniques to new land, but in the 1960's and 1970's the rapid expansion also reflects the increasingly widespread use of plastic vinyl greenhouses to extend the length of the growing season at both ends of the winter. The data on double-cropping, then, reflect both provincial natural endowment in terms of climatic suitability for two crops and the success of human efforts to amplify the effects of that natural endowment. It is worth noting once again that mountain-bound Kangwon province is the only region unable to make use even of vinyl to successfully expand its double-cropped paddy percentage.

A description of South Korea's rural regions would not be complete without mentioning the spatial pattern of the economy's urban industrial growth. Referring back to Map IV-5, one can see that mountainous Kangwon and the southern portion of the rice belt have the least contact with cities, in particular large industrial cities. The hybrid provinces of Kyonggi and South Kyongsang, however, contain not only the very large Special Cities of Seoul and Pusan, but each also has its complement of satellite cities, Suweon and the port of Incheon in Kyonggi and the industrial and port cities of Ulsan and Masan in South Kyongsang. In addition, South Korea's third largest industrial city, Taegu, is in lower North Kyongsang Province and therefore within the orbit of the southeastern hybrid region.

Of the remaining urban industrial concentrations, it is worthwhile mentioning the large city of Taejeon in South Chungchong Province, both because it is a large city in its own right (fifth after Taegu and Incheon), and also because its geographical location is at a large-scale

Table IV-5.   *Manufacturing Employment of Major Industrial Cities, 1966-1973*

(thousands of persons)

City	1966	1968	1970	1973
A. Seoul	180.2	253.5	291.7	383.5
B. Pusan	102.8	129.5	137.3	186.0
C. Incheon	24.5	32.3	47.0	65.2
D. Suweon	4.4	4.9	13.0	13.9
E. Chuncheon	1.9	2.1	5.2	3.2
F. Cheongju	6.1	6.8	6.0	8.4
G. Taejeon	17.7	21.7	23.8	25.4
H. Jeonju	9.4	10.7	11.4	9.9
J. Gunsan	7.4	9.0	10.4	10.6
K. Kwangju	14.7	18.5	17.3	18.8
L. Mogpo	5.7	6.0	5.6	4.7
M. Taegu	52.7	67.9	66.9	75.9
N. Masan	5.4	16.8	23.3	47.4
P. Ulsan	9.0	7.0	10.4	17.6

Source: From Sa-hon Kim, "Kwanggong-eop Senseosu Charyo," [Statistics from Industrial Censuses], Seoul, 1975 (unpublished, in Korean). Firms with fewer than four workers were excluded from the data.

Note: The capitalized letters (A through P) are used to identify these 14 cities on maps such as Map IV-5 on page 87. Thus, for example, Incheon has a border of C's, Suweon a border of D's, Chuncheon a border of E's, and so forth.

"fork in the road" for natural routes of communication between Seoul in the North and the various hinterland provinces in the South. The central region of South Korea, focused on Taejeon and including much of western North Chungchong Province, is of particular importance in the process of Korea's rural response to rapid industrialization, more because of its central location than because of its own immediate industrial intensity.

The relative size of all these important industrial centers can be seen in Table IV-5, "Manufacturing Employment of Major Industrial Cities, 1966-1973," and it is impossible not to remark at once the imposing advantage of Seoul and Pusan in comparison to other cities. It is also significant that the "satellite" cities mentioned above are comparatively large and represent some of the fastest growing industrial areas in the country.

Beginning, then, with the natural pattern of land and climate, the industrial centers of South Korea impose a second matrix of influences on the economy's rural community. We can safely say that in spite of the basic homogeneity of all of South Korea's farmland society, differences imposed by mountains, temperature, rainfall and modern economic growth provide an excellent opportunity to separate and examine those forces which have been responsible for agricultural development registered in but also obscured by the trends in national total statistics examined thus far.

CHAPTER V

REGIONAL FARM PRODUCTION AND GROWTH UNDER JAPANESE COLONIAL CONTROL (1910-1945)

The sweep of rural development in twentieth century Korea reveals that patterns of regional diversity so apparent by 1975 reflected in fact little more than a maturing of evolutionary tendencies originating in the Japanese colonial period. The surging superior productivity of the Northwest around Seoul, the poverty of Kangwon, output's struggle to match population increase in the rice belt regions, and the ascendency of traditionally underutilized "upland" areas—these and other manifestations of South Korea's recent rural modernization are also foci of attention for students of the earlier period. Furthermore, while some early tendencies were indeed only hints of later trends, the dominance of Kyonggi province was undisputed by 1940, and vegetables had clearly made their mark as a major crop category.

The present chapter will analyze this period's 35-year path from early relative equality to considerable regional divergence in 1940. Statistical problems can not disguise the degree of early provincial parity, and yet, within a short few years, changing rice yields and swings in population put Kyonggi well out in front and left Kangwon quite far behind. The causes of this shift are not altogether clear, but chemical fertilizers seem to have played an important role. If, however, rice yields, population change, vegetable growth and all the factors supporting them induced such regional variation, why did they favor Kyonggi? The city of Seoul is the only satisfactory explanation. The spell of this and other urban areas on South Korean rural development is as apparent here as it is during the years of later rapid industrial growth. The mechanism of this relationship is unclear and

Table V-1. Provincial Gross Farm Product per Household, 1920-1940

(thousand 1970 won per household; 3-year average)

	1920 -22	1923 -25	1926 -28	1929 -31	1932 -34	1935 -37	1938 -40
Kyonggi	142	126	128	138	205	238	199
Kangwon	103	97	102	101	109	115	112
N. Chungchong	120	112	117	115	136	157	145
S. Chungchong	147	134	144	129	148	181	142
N. Jolla	117	118	123	123	138	152	141
S. Jolla	122	126	131	118	128	132	129
N. Kyongsang	137	125	117	118	119	137	130
S. Kyongsang	123	121	126	114	125	152	136
All Chosun[a]	127	120	125	124	140	161	150

Sources: Harvest levels for individual crops and individual years were summed using 1970 crop prices; provincial totals were then divided by provincial farm household data from Chosun Government General administrative data. For details and source citations see Appendix A.

Note: [a] This average includes the per-farm outputs of North Korean provinces as well.

demands attention it has not yet received, but the presence and pull of the force it relays is real and hard to ignore.

The performance of Kyonggi Province is the single most remarkable feature of the colonial period, and the degree to which Kyonggi outpaced other areas can be seen in Table V-1, "Provincial Gross Farm Product per Household, 1920-1940." From a second-place position in the 1920's only slightly ahead of several other provinces, Kyonggi began a dramatic increase in its per-farm output during 1929-1931, and by 1940 the northwestern province was producing over a third more output per family than the cluster of second-place provinces. Its performance was unparalleled in either real or percentage terms. The 1938-1940 dip in average product is a result of the nationwide disastrous rice harvest of 1939, and when individual years rather than averages are reviewed, Kyonggi's performance is even more striking.[1] Within the span of twenty years Kyonggi had established the pattern of regional production that would dominate

---

[1] 1940 was also a bad year for rice in the north as well, and the 1938-1940 period saw the start of a decline in vegetable production which became more pronounced as the war continued. For the details of annual data see Appendix A.

South Korea through the 1970's.

The same table shows other interesting performances as well. It is clear that Kangwon Province was already poor by the beginning of the century and that it did not share proportionally in the growth of the 1930's. It is also interesting that the wealthiest province at the beginning of the period, South Chungchong, all but stagnated until after 1935. This province has a long tradition as the home of scholars and aristocracy (the "Yangban" of Confucian Korea)[2] and shares some of the characteristics of hybrid Kyonggi and upland North Chungchong. The sudden reversal in ranking of Kyonggi and South Chungchong is perhaps the single most interesting regional puzzle of the period.

An additional interesting relationship is that between the two ricebowl Jolla provinces. While South Jolla enjoyed a slight lead in the 1920's, it was decisively eclipsed in the 1930's by its northern neighbor. We will see that North Jolla received a disproportionate share of colonial irrigation projects and was remarkable for its concentration of Japanese landlords, while the inferior position of South Jolla first appearing in this period becomes apparent again, and indeed exaggerated, during the years of rapid growth in the 1960's.

Finally, giving a preview of its post World War II meteoric rise in per-household and per capital product, North Chungchong Province, typifying the upland region with some access to towns, was actually in second place by 1938-1940. North Chungchong in the 1960's displayed remarkable increases in both vegetables and rice to supplement its dominant traditional position in the tobacco industry, and although its growth in the 1930's was nowhere nearly as precocious as Kyonggi's, it is clear that some of the forces responsible for its much later successes were by then already working.

The period from 1920 to 1940, then, was clearly one of rapid development and growing regional divergence. What was the nature of the initial equality among the provinces, and what were the components together accounting for later imbalances? The passages which follow will address these individual issues.

---

[2] A 1910 Japanese survey of population by occupation found a full 10 percent of South Chungchong to be Yangban and their families, while the percentage for Kyonggi and the country overall was less than 3 percent. See Zenski Eitsuke, "Chosun ui Ingu Tongei" [Chosun Population Statistics] in Bureau of Statistics, Economic Planning Board, *Hangug ui Ingu Dongtae Tongei* [Vital Statistics of Korea], Seoul, 1965. (in Korean)

The establishment of original rough equality between the provinces of South Korea is not at all straightforward. Systematic reporting of provincial crop levels did not begin until 1910, and for some crop categories (vegetables and potatoes) not until 1912. Furthermore, coverage for individual crops within crop categories was far from complete in the first decade of reporting and even later.[3] By far the most serious difficulty, however, results from the apparent poor quality of administrative reporting before 1918.

By the time of the completion of the cadastral land survey in 1918, it had become clear to Japanese officials that estimates of crop yields and crop harvests were seriously low.[4] Special reporting teams were sent to all parts of the country, and their data on the harvests of 1918 together with information from the just-completed land survey were used for a wholly new set of crop yield and output estimates. Furthermore, on the basis of the new results, all data from 1910 to 1917 were revised upward and thenceforth published at the new levels. The revisions were substantial. For rice, output for the years 1910 to 1917 were increased between 10 and 30 percent. For barley the adjustments were between 30 and 90 percent. Available data on these adjustments, however, are of little direct use for regional analysis, since only national aggregate adjustments were tabulated. If the discrepancies in reporting were uniform from one region to the next, the original provincial data, although low in an absolute sense, would nevertheless convey an accurate picture of relative provincial productivity.

Table V-2, "Provincial Farm Product per Household, Individual Years, 1912-1918," has been calculated from provincial crop levels as originally reported. The jump in reported output between 1917 and 1918 is evident, but the jump is not uniform across all provinces. There are several reasons for believing that the under-reporting of the 1910-1917 years was, indeed, not uniform. In the first place, examination of the level of adjustments for all crops and for individual categories reveals that those for Kyonggi, South Kyongsang and South Chungchong were relatively small (less than 20 percent), while for other more isolated provinces the adjustments were in fact much larger (as much as 50 percent for Kangwon and North Kyongsang). It

---

[3] For exact information by crop category on the degree of coverage by year from 1910 to 1940 see Appendix A.

[4] This and the account which follows relies heavily on text and tables in Kuro Komikawa, *Chosun Nogyo Hattatsu-shi* [History of Chosun Agriculture], Volume 1 (Development Volume), Seoul, 1944, (in Japanese), pp. 137-143.

is understandable that lands closer to administrative centers such as Seoul, Pusan and Taejeon should have somewhat more careful reporting. Secondly, examination of the 1917-1918 jump for crop categories shows that it was much smaller for rice than for other crops.[5] This is also reasonable when the importance of rice for taxes and tenant payments is remembered. In other words, provinces with a heavier dependence on rice could be expected to have more accurate overall crop reportage than could regions with a smaller share of total output in rice. Finally, the cadastral survey was not carried out simultaneously in all provinces. As reports from provinces surveyed early became available after 1911, it is reasonable to assume they were used to calculate output estimates more accurately than those from provinces relying on pre-survey data.

It is true that the data from before 1918 show greater inequality between provinces than those for 1918 and after. The above analysis has argued, however, that the scale of the differences was exaggerated by regional discrepancies in reporting accuracy, and hence it is likely that the most accurate picture of the degree of regional equality at the beginning of Korea's twentieth century can only be found in data from the 1920's. The attention paid to this problem has not been too great. Was per-household production at the end of the Yi Dynasty in fact in a state of equilibrium? Analysis of available data indicates that it was. The inequality of the 1930's can then be viewed as the result of forces disturbing that equilibrium. What was the equilibrium's nature? Were the components of per-household output also comparable between provinces? The analysis which follows will show that greater output per farm hectare was in fact compensated for by greater farm population density in areas where the land could support larger numbers.

The single most significant difference between provinces in this early period was the much greater farm population density in the South. Table V-3, "1927 Provincial Farm Size," shows that while there was more than a hectare and a half for households in Kyonggi, there was less than a hectare per family in South Kyongsang, the most densely populated region. The same table shows that though the difference was less extreme than for upland, Kyonggi and South Chungchong also had much higher ratios of paddy to farm households than

---

[5] See the Annual Provincial Series for individual crop categories, 1910-1940, in Appendix A.

Table V-2. *Provincial Farm Product per Household, Individual Years, 1912-1918*

					(thousand 1970 won per household)		
	1912	1913	1914	1915	1916	1917	1918
Kyonggi	89	95	107	118	121	120	143
Kangwon	66	65	74	71	68	73	110
N. Chungchong	58	59	83	87	86	93	121
S. Chungchong	83	85	113	104	129	125	147
N. Jolla	72	85	104	93	95	101	117
S. Jolla	67	77	86	91	88	95	118
N. Kyongsang	73	75	85	76	85	91	138
S. Kyongsang	78	78	94	94	97	107	122
All Chosun[a]	79	81	90	90	96	97	128

Source: Calculated from individual province crop levels, 1970 prices for each crop and administrative data on provincial farm households. For details see Appendix A.

Note: [a] These totals include data from provinces in North Korea as well.

areas further south. Kangwon, of course, is the least-well endowed of all provinces in this respect, and the extremely small size of the average family holding helps begin to explain the region's inferior position throughout the century. It is also important to notice that although families in the North are larger than in the South, the difference (8 percent between Kyonggi and South Jolla) is not significant when compared to the difference in land holding (47 percent for paddy and 42 percent overall). It is interesting that this severe regional imbalance in farm population density was paralleled by comparable differences in rural wages at the time. Early Japanese surveys before annexation reveal that wages from numerous counties in Kyonggi and South Chungchong were all between two and three times the level reported in over two dozen counties in the Jolla and Kyongsang provinces.[6]

If the regional skew in population density was severe, it was balanced to a considerable degree by greater productivity of southern soil. This is most clearly seen in the first column of Table V-4,

---

[6] The wages were for a day's work when meals were supplied by the farmer. Similar data on the level of annual reimbursement for a live-in year round hand show the same degree of difference between north and south. See Sei Yamaguchi (ed.), *Chosun Sangyoshi* [Chosun Industrial Report], Tokyo, 1910, Volume I, (in Japanese), pp. 335-344.

Table V-3. 1927 Provincial Farm Size

	Number of Households (thousand)	Persons Per Household	Paddy per Household (100 ares)	Upland per Household (100 ares)	Land per Household (100 ares)
Kyonggi	236.9	5.4	85.3	78.3	163.6
Kangwon	199.4	5.5	42.7	127.8	170.5
N. Chungchong	133.1	5.3	52.3	66.9	119.1
S. Chungchong	180.0	5.4	89.0	46.2	135.1
N. Jolla	215.9	5.1	77.9	31.3	109.2
S. Jolla	350.4	5.0	58.0	56.8	114.8
N. Kyongsang	337.4	5.4	56.4	59.1	115.5
S. Kyongsang	295.0	5.2	58.0	36.6	94.6
S. Korea	1,948.1	5.3	64.2	60.9	105.2

Source: Calculated from Chosun Government General, *Tokei Nenbo, Showa Ninen* [Statistical Yearbook, 1927], Seoul, 1929. (in Japanese)

"Provincial Household Output for Major Crop Categories, 1927," which gives the level of rice output per family for each province. It is of course striking, that in contrast to the small paddy allotment per southern farm, rice output per household there matches and in many cases surpasses that in Kyonggi. Perhaps equally important is the variation in output of barleys, the other major foodgrain category for South Korea. Barley is grown in two ways in South Korea, either as a summer crop on hillier upland fields or as a winter crop on paddy after the rice harvest is in. As we have seen in the previous chapter, this second method, so-called "double cropping," had always been much more prevalent in the warmer southern provinces. This in large part accounts for the greater output of barley in South Jolla and South Kyongsang, and it is not difficult to envision how this process of double cropping in fact increases the effective land endowment of southern households, thereby aiding in the process of compensation for nominally smaller farms.

In discussing the three other major crop categories, pulses, special crops and vegetables, it is easiest to look at Kyonggi and South Jolla for comparisons. Although South Jolla households produced more foodgrains than those in Kyonggi, Kyonggi outproduced the southwestern region in crops suited to upland, most notably beans, but also vegetables. This northern specialization in pulses was itself compensated for in value terms by South Jolla's production of special crops,

Table V-4.   Provincial Household Output for Major Crop Categories, 1927

(thousand 1970 won per household; 5-year average)

	Rice	Barleys	Pulses	Special Crops	Vege-tables	All Crops
Kyonggi	72.8	9.6	22.9	4.7	13.0	128.1
Kangwon	41.4	7.1	22.5	6.8	7.9	101.7
N. Chungchong	50.3	14.2	17.5	23.7	5.4	115.4
S. Chungchong	86.6	11.5	18.9	11.1	10.2	140.1
N. Jolla	87.8	9.4	9.1	8.9	5.1	121.8
S. Jolla	69.6	17.0	7.3	23.0	4.0	127.4
N. Kyongsang	59.4	19.3	18.0	7.9	5.6	114.9
S. Kyongsang	70.2	18.1	10.7	10.5	3.9	116.1

Sources: Calculated from 1970 crop prices, individual crop output levels for 1925-29, and province farm household data. For details see Appendix A.

in this case almost totally composed of cotton. The comparison of these two provinces and the different structure of their nevertheless comparable household products is seen in Table V-5, "Major Crop Shares in Provincial Household Output, 1927," which translates Table V-4 into percentage terms. The point is, of course, that the sum effect of different specialization patterns and productivity levels was to equalize overall product per household in the two regions. It is true that Kyonggi farms were already more productive than southern farms in 1927, but as we have seen in Table V-1 the margin is small in comparison to the differences which later emerged, and subsequent pages and a table for 1935 matching Table V-4 will help to accent important components of Kyonggi's rapid emergence.

Relative rice yields are of course crucial to an understanding of the regional balance in rural Korea of the early twentieth century. Table V-6, "Provincial Rice Yields, 1923-1938," shows not only this relationship for the 1920's, but also the degree to which it was disturbed in the following decade. Looking at the average yield for the five years centered on 1923, the yields for the rice belt and southern provinces are over a third greater than for Kyonggi. When allowance is made for the added productivity of double-cropped barleys, this advantage nearly makes up for the lower paddy endowment documented in Table V-3. This compensating greater southern rice productivity was maintained through 1928, but by 1933 Kyonggi's yields had all

Table V-5.   *Major Crop Shares in Provincial Household Output, 1927*

(%; 5-year average)

	Rice	Barleys	Pulses	Special Crops	Vege-tables	Other
Kyonggi	56.4	7.5	17.9	3.6	10.1	4.5
Kangwon	40.7	7.0	22.0	6.6	7.8	15.9
N. Chungchong	43.4	12.3	15.1	20.7	4.6	3.9
S. Chungchong	61.6	8.1	13.4	8.1	7.2	1.6
N. Jolla	71.7	7.8	7.4	7.4	4.2	1.5
S. Jolla	54.5	13.4	5.7	18.1	3.1	5.2
N. Kyongsang	51.1	17.1	15.4	7.0	5.0	4.4
S. Kyongsang	60.0	15.9	9.1	9.1	3.3	2.6

Sources: Calculated as the average of shares for each category from 1925-29. For discussion see note to Table V-4 and Appendix A.

Table V-6.   *Provincial Rice Yields, 1923-1938*

(tons per hectare; 5-year average)

	1923	1928	1933	1938	Annual Percentage Change 1923-38
Kyonggi	1.13	1.18	1.43	1.57	2.2
Kangwon	1.23	1.36	1.42	1.73	2.3
N. Chungchong	1.34	1.29	1.37	1.64	1.4
S. Chungchong	1.30	1.31	1.29	1.65	1.6
N. Jolla	1.58	1.45	1.46	1.66	0.3
S. Jolla	1.52	1.59	1.46	1.67	0.6
N. Kyongsang	1.58	1.45	1.46	1.66	0.3
S. Kyongsang	1.60	1.63	1.52	1.74	0.6

Sources: Calculated from provincial rice output and paddy data in Appendix A.

but pulled even with those in the South, and throughout the 1930's the southern provinces were unable to regain their former superiority in this respect. With more equal rice yields, of course, the difference in paddy endowment becomes extremely significant, and the more rapid worsening of the land-man ratio in the South between the 1920's and 1930's is made just that much more important in the overall evolution of regional inequality.

For although the area of cultivated land showed little increase during this period,[7] rates of farm population change were very different between provinces and from one period to the next. Table V-7, "Provincial Farm Household Growth, 1910-1941," shows the average annual rate of farm household growth in each province for the 1920's and 1930's. The most interesting period for comparison is that between 1926 and 1935. While Kyonggi's total number of farm households held steady (. 2 percent per annum growth), that for South Chungchong, for example, exploded (2.3 percent per year). The two Jolla provinces also showed greater-than-average growth during both this and the following set of years, and overall the rice belt provinces between 1926 and 1941 registered increases in total farm population substantially higher than elsewhere, with the exception of Kangwon.[8] The net effect of this differential farm population growth was to accentuate further the bias in land endowment favoring Kyonggi.

The demographic forces responsible for the above trends are most certainly complex, but the low farm population growth in Kyonggi, South Kyongsang and North Kyongsang, with their urban centers (Seoul, Pusan and Taegu), is perhaps a sign that demand for labor in the cities was attracting families out of rural communities. It is possible that contact with cities in some way changed birth rates, and the differential origins of those who migrated to Manchuria and Japan could also have been important.[9] A demographic analysis of this

---

[7] For annual provincial series of paddy and upland from 1910 to 1942 see Young-il Chung, "Kyeong-chi Myeon-cheok ui Chu-kye wa Bun-seok (1911-1971)" [Estimation and Analysis of Arable Land Area (1911-1971)], *The Korean Economic Journal,* June 1975. (in Korean)

[8] The data are for households rather than population and hence it is possible that changing household size could make up for the apparent differences shown in Table V-7. After the 19-teens the Japanese stopped reporting farm population by province, but farm families were two-thirds of all households in South Korea with higher shares in rice-belt and non-urban provinces, and data on average size of all households by province shows little variation in 1941. In particular, Kyonggi's farm household size (5.5 persons) matched exactly that of South Chungchong, the most rapidly growing province. South Jolla and South Kyongsang did have smaller households (5.2 and 5.3 persons) but the difference is insignificant compared to the variation in household growth rates. The 1941 Data are from the Chosun Government General, *1941 Statistical Yearbook,* Seoul, 1942, p. 26.

[9] In fact, one study reports that migration both to Japan and to Manchuria in the 1930's originated in the south, without, however, differentiation between South Jolla and South Kyongsang. See Tai-hwan Kwon, et al., *The Population of Korea,* Seoul: Seoul National University, 1975, pp. 28-29.

Table V-7.   Provincial Farm Household Growth, 1910-1941

(%; annual average)

	1910-1926	1926-1935	1935-1941	1926-1941	1910-1941
Kyonggi	1.0	.2	.5	.3	.7
Kangwon	2.3	2.3	− .4	1.2	1.7
N. Chungchong	1.6	.8	− 1.0	.1	.9
S. Chungchong	.2	2.3	1.0	1.8	1.0
N. Jolla	1.4	1.3	.6	1.0	1.2
S. Jolla	2.2	1.4	1.3	1.3	1.8
N. Kyongsang	1.1	.9	− .7	.3	.7
S. Kyongsang	.8	.5	.2	.3	.6
S. Korea	1.3	1.2	.3	.8	1.1

Source: Calculated from Chosun Government General Administrative data. For data and sources see Appendix A.

phenomenon is beyond the capacity of this study, but its necessary impact on average farm productivity must be clear.

The combined force of changes in rice yields and differential farm population change resulted in dramatic reversals in the ranking of provinces by per-household rice productivity. Table V-8, "Provincial Rice Output per Household, 1920-1940," makes it very clear that between the 1920's and 1930's Kyonggi came from a position lagging the rice belt and the South (notice in particular the ranking for 1926-1928) to one of clear superiority. Throughout the years 1932-1940 Kyonggi increased and maintained its new lead. Indeed, Kyonggi was to keep its position throughout most of the post World War II period and show a similar and even greater surge in the 1960's and early 1970's.

A complex of patterns, then, influenced regional change in rice productivity, and this trend alone goes a long way towards describing Kyonggi's overall superior performance. Other crop categories, however, and in particular vegetables, were also important. Analysis of Table V-9, "Provincial Household Output for Major Crop Cetegories, 1935," will show how this is true and will complete our review of rural South Korea's development away from the century's early rough regional equality.

Table V-9 provides the same information for 1935 that Table V-4 gave for 1927. The difference after a short eight-year period is striking. Output is in general higher everywhere in 1935 than in 1927, but

Table V-8.   *Provincial Rice Output per Household, 1920-1940*

(thousand 1970 won per household; 3-year average)

	1920 -22	1923 -25	1926 -28	1929 -31	1932 -34	1935 -37	1938 -40
Kyonggi	76	71	73	82	98	113	93
Kangwon	39	38	42	43	41	49	50
N. Chungchong	57	53	51	51	54	67	59
S. Chungchong	100	86	90	80	77	105	78
N. Jolla	85	86	89	91	88	94	88
S. Jolla	68	70	73	64	67	65	64
N. Kyongsang	75	67	61	65	59	75	65
S. Kyongsang	70	72	80	69	66	90	75

Sources: Averaged from annual data compiled from individual harvest levels, the average 1970 rice price and administrative farm household data. For details and annual data see Appendix A.

Table V-9.   *Provincial Household Output for Major Crop Categories, 1935*

(thousand 1970 won per household; 5-year average)

	Rice	Barleys	Pulses	Special Crops	Vege- tables	All Crops
Kyonggi	108	14	23	7	67	227
Kangwon	46	6	19	6	19	111
N. Chungchong	62	16	17	23	25	148
S. Chungchong	95	14	15	12	30	169
N. Jolla	94	14	9	14	20	182
S. Jolla	65	22	5	19	11	129
N. Kyongsang	70	20	15	13	10	130
S. Kyongsang	80	23	7	16	11	141

Sources: Calculated from 1970 crop prices, individual crop output levels for 1933-37 and administrative farm household data. For details see Appendix A.

the increase for Kyonggi is greatly out of scale with that for the other provinces. Looking at the shift in provincial output for crop categories, the most striking by far is that of vegetables. The per-farm output for barleys, pulses and special crops retained their rough relative values, but vegetables displayed a tremendous increase in production, an increase, moreover, heavily skewed in favor of Kyonggi. If we temporarily focus on Kyonggi and South Chungchong, the latter province most dramatically eclipsed by the former, we notice that where Kyonggi's overall product lagged behind that of South Chungchong by 12,000 won in 1927, by 1935, Kyonggi had produced a lead of 58,000 won, a net gain of 70,000 won. A sizable portion, 27,000 won, of this increase was due to Kyonggi's increased productivity of rice, but a full 40,000 won was due to vegetable production.

It is important to say at once that a large proportion of this difference is a statistical creation, and this fact presents our analysis with considerable difficulties. The coverage of Japanese colonial crop data was constantly changing as new crops were included, but in 1932 coverage for vegetables was drastically extended from 2 crops to 12. For all of Korea this represented an increase of well over 200 percent in reported vegetable production and for Kyonggi an increase of over 300 percent. The two crops reported since 1912, cabbage and white radish, are indeed the most important by volume for the Korean diet, but the expanded coverage in the thirties showed that these were overshadowed in 1970-won value terms by the combination of other more "exotic" vegetables.

The crucial question, of course, is the degree to which these same exotic vegetables had held a comparable position in total output earlier in the century. If they had, and if Kyonggi's relative share had also been as large, it would be much more difficult to argue that there had been rough equality in household output between South Korea's regions. It is difficult to believe, however, that secondary vegetables were so important in the 1920's and before. For one thing, the decision to extend coverage in 1932 must have in some way been related to an official realization that these crops were more important then they had been thought to be in former years. This was perhaps the result of faulty perceptions during the earlier period, but it also is likely that vegetables had in fact become more important than they had been previously. In the second place, it is safe to say that the increase in vegetable output, and in particular the increase in exotic vegetables, was related to the growth of urban markets. Between 1915 and 1935 the combined population of Seoul and its port of Incheon grew by over 200,000 persons, more than a 75 percent increase. The industrial and port city of Pusan increased during the same period by 120,000 persons, a jump of almost 200 percent.[10] It is unlikely that secondary

---

[10] The Data are from Chosun Government General, *Statistical Yearbook,* relevant years.

vegetable production was as important before this rapid process of urban concentration. Finally, and in support of the point just made, a study of the regional growth of individual secondary vegetables after 1932 shows that while their level of output remained roughly constant in the rice belt and more isolated provinces, it grew rapidly and steadily in Kyonggi and South Kyongsang provinces until 1938, after which there was a nationwide decline in vegetable production which lasted until after the Second World War.[11] This growth after 1932 in precisely those provinces where urban centers had been growing and were continuing to grow strongly suggests that similar regionally differentiated growth had been going on prior to 1932, and that, by extension, the regional differences in output of "exotic" secondary vegetables were much less pronounced in the early years of the century, if they existed at all.

It seems fairly clear, then, that the contribution of vegetables to regional inequality in the 1930's had been a relatively recent phenomenon. Comparable reporting in the 1920's may have shown an earlier point of emergence for Kyonggi Province, but this chapter's basic conclusions remain unaffected. If data we have used for the 1920's are accurate indicators of relative conditions at the beginning of Japan's colonial control, their comparison with available data for the 1930's is a fair test of trends in South Korea's regional rural growth.

Kyonggi Province clearly and decisively broke away from the pack. The preceding passages have detailed the yield, crop and population patterns responsible for its success, but what, in turn, were the forces responsible for these component changes? Were there identifiable Japanese policies which favored Kyonggi? The only conclusion possible from available data is that there were not. Provincial data on the quantity and quality of agricultural inputs are few, but those

---

[11] The data show growth for Kyonggi of over 60 percent in value terms for all vegetables between 1932 and 1938, while comparable growth for all of Chosun, including Kyonggi and South Kyongsang, was only 18 percent. For secondary crops only, measured in metric tons, gross output for Kyonggi also increased 60 percent as compared to only 13 percent for all of Chosun. The provincial series for individual crop output from 1910 to 1940 were converted to tons from Chosun Government General data in Korean weight and volume units as compiled by Shigeru Ishikawa in *Chosun Nogyo Seisankaku no Shukei, Senzen no Bu* [Collected Statistics of Chosun Agricultural Production, Pre-war Part], Economic Research Center, Hitotsubashi University, Tokyo, 1973, (mimeographed, in Japanese). A projected post-war part was never compiled.

documenting the intense Japanese effort at land improvement in the late 1920's show that Kyonggi was not particularly favored. The paragraphs which follow will outline what information there is about improved cultivation conditions in pre-World War II South Korea.

We have already seen that the 1920's were for the Japanese a period of concerted effort to increase rice output for export to Japan. This effort was organized under the "Plan to Increase Rice Production" of 1920, and an important part of that plan was the longterm "Land Improvement Program" begun in 1925 after the initial phase of a ten-year feasibility survey started in 1920[12] The land improvement program was to have lasted 16 years, until 1940, but because of the Depression and rice riots in Japan it was prematurely terminated in 1931 with much of its originally planned work left undone. Nevertheless, over 10 percent of South Korea's arable land was affected by the program in its brief six-year history, and it is interesting to notice the regional distribution of these land improvement works.

There were four basic kinds of land improvement: irrigation, conversion of dry fields to paddy, the clearing of new land, and reclamation of land from the sea. Irrigation accounted for by far the larger share of work actually completed by 1931 (57 percent), and Japanese statistics show that irrigation brought about a doubling of yields for irrigation association paddy.[13] The degree of success shown by such before-and-after comparisons is undoubtedly exaggerated, both because irrigation association land was probably better land and because other changes, most notably in seeds and fertilizer, were being introduced at the same time. Nevertheless, the importance of secure irrigation for dependable high yields cannot be questioned.

Table V-10, "Provincial Irrigation Work, 1914-1931," gives the area of paddy land in each province benefited by irrigation work, both before and after the inception of the land improvement program. It is clear that the land improvement program in fact only formalized a process which had been going on for some years. The most striking feature of Table V-10 is how North Jolla Province received an overwhelmingly large share of irrigation projects both before and after the program's inception. This fact is largely explained by the heavy concentration of Japanese landlords in this province and

---

[12] For an account of these plans and programs, see Chosun Government General, *Tochi Kairyo Jigyo no Gaikyo* [The General State of Land Improvement Works], Seoul, 1932, (in Japanese), pp. 1-8.

[13] *Ibid.*, Appendix, pp. 14-16.

Table V-10.   Provincial Irrigation Work, 1914-1931

(thousand hectares)

	Before 1925	After 1925	Total	Share of Paddy (%)		
				Before 1925	After 1925	Total
Kyonggi	6.3	8.8	15.1	2.1	4.4	7.5
Kangwon	10.2	5.5	15.7	12.2	6.6	18.7
N. Chungchong	1.0	1.5	2.4	1.4	2.1	3.5
S. Chungchong	4.1	5.4	9.5	2.5	3.3	5.9
N. Jolla	20.0	20.4	40.5	11.6	11.9	23.5
S. Jolla	3.9	5.3	9.2	1.9	2.6	4.5
N. Kyongsang	2.5	5.0	7.5	1.3	2.6	3.9
S. Kyongsang	14.3	6.3	20.7	8.4	3.7	12.2
Total	62.3	58.3	120.6	5.0	4.6	9.6

Source: Calculated from data in Chosun Government General, *Tochi Kairyo Jigyo no Gaikyo* [The General State of Land Improvement Works], Seoul, 1932, Appendix, (in Japanese), pp. 19-39. The table reports irrigation projects both under the auspices of and independent of the Japanese-sponsored irrigation societies.

their considerable success in the reclamation of paddy from coastal areas. The important conclusion, of course, is that Kyonggi Province was not particularly favored and even trailed South Kyongsang.

Irrigation, however, did not begin with the Japanese, and it is valuable to look at the overall status of irrigation by province in the colonial period, rather than just the additions made after 1910. Table V-11, "Overall Status of Irrigation by Province, 1932," shows that of all the provinces Kyonggi had the lowest share of its paddy securely irrigated and the highest share dependent on rainfall alone. While South Korea in 1932 had an average of 42 percent of its land safely irrigated, Kyonggi had only 30 percent, and while the average for all provinces was 36 percent for rain-fed paddy, that for Kyonggi was 46 percent. In other words, neither Kyonggi's overall pattern of irrigation nor the changes made in water supply under the Japanese gave it an advantage that can be translated into the yields and output increases of the 1930's.

There were, however, other factors affecting output during the 1920's and '30's, and the remaining paragraphs will review the importance of new seed varieties, chemical fertilizer and patterns of tenancy and land ownership.

Table V-11.   Overall Status of Irrigation by Province, 1932

(thousand hectares)

	Securely Irrigated	Insecurely Irrigated	Rain-fed	Securely Irrigated	Insecurely Irrigated	Rain-fed
				Share of Total (%)		
Kyonggi	61.1	50.8	94.8	29.6	24.6	45.9
Kangwon	61.4	12.1	16.8	68.0	13.4	18.6
N. Chungchong	40.0	14.5	17.9	55.2	20.0	24.7
S. Chungchong	49.1	39.4	74.8	30.0	24.1	45.8
N. Jolla	85.7	28.9	58.7	52.2	17.6	35.7
S. Jolla	80.8	50.6	78.6	38.5	24.1	37.4
N. Kyongsang	90.7	48.7	56.7	46.3	24.8	28.9
S. Kyongsang	75.6	36.3	65.8	42.5	20.4	37.0
Total	544.4	281.3	464.1	42.2	21.8	36.0

Source: Converted to metric units from data in Ichishi Ogawa, Seiichi Higashihata, *Chosen Maikoku Keizairon* [The Economics of Chosun Rice], Tokyo: Japanese Technical Society, 1935, (in Japanese), p. 47.

Agricultural experimentation and extension education was begun by the Japanese even before annexation in 1910, but its growth in early years was slow.[14] The center of this activity was in Suweon, to the South of Seoul, where the only agricultural secondary school in Korea was founded in 1906 and where over the decades a system of laboratories and experimental fields was developed and expanded. The work eventually came to concentrate on new seed varieties, pest and disease control and better implements and techniques for the production of rice and other major crops.

One of the most significant achievements of this work was the development of rice seed varieties from Japanese and superior Korean breeds and their dissemination throughout the country. At the beginning of the Japanese period there was a dizzying array of traditional varieties, many of them quite similar and yet going by different names and others very different biologically, yet classified by Korean farmers as one type.[15] The varieties developed by the Japanese

---

[14] This account is taken from Yun-seong Hwang, *Hangug Nongeop Kyoyuk Sa* [A History of Agricultural Education in Korea], Seoul: Taehan Chulpan Sa, 1964, (in Korean), pp. 133-138.

[15] See Choji Hishimoto, *Chosun Mai no Kenkyu* [A Study of Chosun Rice], Tokyo, 1938, (in Japanese), pp. 142-143.

Table V-12. Use and Yield of New Rice Seed Varieties, 1912-1936

	Share of Paddy Planted to New Varieties (%)	Share of Harvest from New Varieties (%)	Yield from New Varieties (tons/ha.)	Yield from Traditional Varieties (tons/ha.)
1912	2.8	4.6	1.84	1.09
1917	39.1	51.4	1.71	1.05
1922	63.6	70.6	1.55	1.13
1927	74.1	79.4	1.68	1.26
1932	77.6	80.1	1.51	1.29
1936	85.3	86.8	1.80	1.60

Source: Calculated from Kuro Komikawa, *Chosen Nogyo Hattatsu Shi* [History of Chosun Agriculture], Development Volume, Seoul, 1944, (in Japanese), p. 217.

brought higher yields, and their success in replacing the traditional varieties with new ones can be seen in Table V-12, "Use and Yield of New Rice Seed Varieties, 1912-1936." The new seeds were planted to less than 3 percent of all land and accounted for less than 5 percent of all harvests in 1912, but by 1936 the share was greater than 85 percent for both measures. The table also shows that the new seeds brought better yields, but decreasingly so as the years went by, probably because land most suited to new varieties was converted first.

The important thing to notice for our analysis of regional trends is that the bulk of new seed dissemination had already taken place by 1927, still several years before the clear-cut increase in Kyonggi's relative yields. Furthermore, although there are no dependable data on the regional use of new varieties, several surveys of the use of a few major varieties for Korea's provinces, in 1937, show Kyonggi with somewhat larger acreage planted,[16] but hardly enough to account for Kyonggi's exaggerated yield increase relative to other provinces. We are not trying to explain, however, yields in Kyonggi superior to those further south; they were never superior. We need only to understand the forces which made yields roughly comparable, because the higher per-household product was due to the province's larger average farm size. It is quite conceivable that there was less potential for increase in

---

[16] *Ibid.*, pp. 155-163.

Table V-13. *Chemical Fertilizer Consumption and Supply, 1910-1936*

(thousand metric tons)

Year	Consumption	Domestic Production	Net Import[a]
1910	—	—	− 1.1
1911	—	—	− 1.9
1912	—	—	− 5.8
1913	—	—	− 4.3
1914	—	—	− 12.6
1915	—	—	− 15.2
1916	.5	.4	− 15.3
1917	.4	.9	− 22.1
1918	1.5	1.6	− 17.4
1919	3.0	3.6	− 18.2
1920	1.7	2.6	− 19.6
1921	5.5	4.7	− 12.7
1922	4.4	3.1	− 2.2
1923	6.9	5.0	4.3
1924	12.9	1.9	− 2.4
1925	21.3	1.5	1.3
1926	21.5	1.6	87.2
1927	47.6	1.5	112.5
1928	97.0	1.6	140.2
1929	133.3	2.1	154.1
1930	174.7	110.7	82.5
1931	148.2	247.2	20.3
1932	232.4	258.4	− 83.3
1933	289.4	287.1	− 23.9
1934	364.3	—	− 136.7
1935	504.1	—	− 123.4
1936	522.4	—	− 212.8

Sources: Consumption and net imports were calculated from Kuro Komikawa, *Chosen Nogyo Hattatsu-shi,* cited above, Appendix Tables 18, 22 and 23. Domestic Production is from Chosen Sotoku-fu, *Nogyo Tokei Hyo* [Agricultural Statistical Tables], Seoul, 1934, (in Japanese), pp. 107-108.

Note: [a] Traded fertilizer is so-called "Commercial Fertilizer" and includes animal and plant bi-products; this column, therefore, is not comparable with the other two, which refer only to chemical fertilizers.

the heavily planted and already climatically favored rice belt regions. In that case, the combination of land improvements and new seed varieties could have been significant, for Kyonggi, without appearing so in the regional application data.

Of all the important factors explaining farm output, however, perhaps the one with the greatest potential for explaining South Korea's regional pattern of growth under the Japanese is chemical fertilizers. The Japanese did not keep provincial statistics on chemical fertilizer consumption in the 1930's, but the national trends in consumption, domestic production and net import of chemical and commercial fertilizers can be seen in Table V-13, "Chemical Fertilizer Consumption and Supply, 1910-1936." The timing of the extremely rapid increase in consumption of chemical fertilizers between 1925 and 1935 and the increase in Kyonggi's overall productivity per household cannot be dismissed as coincidental. What is more, just as Kyonggi's output spurt begins after 1929, so 1930 is the first year of large-scale domestic production of nitrogen fertilizers. If Kyonggi's consumption of chemical fertilizer was higher than that per hectare in other provinces, or if comparable levels of consumption in Kyonggi nevertheless brought out its greater potential for improvement, the national trends in such consumption might contain a clue about the secret of Kyonggi's rice production effort.

What provincial data do exist for fertilizer consumption cover only the war years of 1941 to 1944 and are presented in Table V-14 in the form of per-hectare of cultivated land consumption levels. If paddy land is used as a base for comparison rather than all cultivated land, the result is not greatly altered, except for Kangwon.[17] The table shows that in yet another important category Kyonggi was less well-favored than the rice belt and southern provinces, in particular North Jolla. Nevertheless, it is still likely that chemical fertilizer played the most important role in raising Kyonggi's per-hectare output of rice to a level on par with the other regions. For one thing, the paddy in southern areas had to do double duty and supported barley crops as well. It is also likely that there were diminishing

---

[17] Separate data for per-hectare fertilizer application to seedbeds and then to rice paddies after transplanting show provincial patterns comparable to those given in Table V-14 for all cultivated land. The data are for 1944 and can be found in a curious and yet excellent little book: Jo-tai Kim, *Chosun Michak Yeongu* [Research in Chosun Rice], Seoul: Jeong-in Sa, 1948, (in Korean), pp. 279-281. The same book contains rice yields by county and *Myeon* [district] for 1941. See pp. 302-327.

Table V-14.   Chemical Fertilizer Consumption by Province, 1941-1944

(kilograms per hectare of cultivated land)

	1941	1942	1943	1944
Kyonggi	31.7	29.2	23.9	18.8
Kangwon	11.8	11.2	9.8	8.5
N. Chungchong	29.9	25.6	25.3	18.9
S. Chungchong	38.6	35.5	29.4	24.1
N. Jolla	55.0	48.6	41.3	27.1
S. Jolla	39.7	34.3	29.2	19.8
N. Kyongsang	33.7	29.1	26.4	18.4
S. Kyongsang	45.4	39.7	32.9	28.2

Source: Calculated from land data in Young-il Chung, cited above and Chosun Bank Research Division, *Chosun Kyeongje Yeonbo, 1948,* Seoul, 1948, pp. 1-362, 363.
Note: The data are for the weight of nitrogen, phosphorous and potash nutrient elements only.

returns to the application of chemical fertilizers, even at these pre-1945 levels. If that was the case, the 31 kilograms per hectare used in Kyonggi in 1941 may have had an effect nearly equal to that of the 39 kilograms in South Chungchong, though perhaps still not comparable to levels in North Jolla and South Kyongsang.

In any event, although the great increase in Korea's overall chemical fertilizer consumption seems suspiciously correlated with the divergence in regional farm product, our analysis of the traditional sources of growth in agricultural output has yielded no clear-cut answer to the riddle of the Northwest's sudden and dramatic performance. The remaining possible factor we will analyze is the regional pattern of tenancy and landlord ownership, which again, however, provides no straightforward answer to our problem.

Although the Japanese eradicated slavery, formalized rent relationships and legalized land ownership, agricultural tenancy in Korea was inherited from the Yi Dynasty and had been the result of centuries of gradual evolution. There were owner-cultivators who tilled their own land or a combination of owned land and rented land, but roughly two-thirds of Korean farmland was rented land, and a similar share of all Korean farmers was made up of tenants.[18] What is more,

---

[18] Japanese data on the incidence of tenancy are abundant. The shares cited in the text are representative of tenancy levels throughout the Japanese period, though there

Table V-15.   1927 Shares of Tenant Land by Province

(percent of provincial total)

	Paddy	Upland	Cultivated Land
Kyonggi	72.8	63.5	68.3
Kangwon	48.4	31.3	35.6
N. Chungchong	62.3	54.2	57.7
S. Chungchong	71.5	49.7	64.0
N. Jolla	78.8	56.4	72.4
S. Jolla	61.1	41.8	47.1
N. Kyongsang	56.2	49.2	52.6
S. Kyongsang	76.0	52.2	59.8

Source : Calculated from Chosun Government General, *Chosen no Sosaku Kanshu* [The Practice of Tenancy in Chosun], Seoul 1928, (in Japanese), pp. 8-9.

although tenancy levels were high throughout the country, they were highest in the rice paddy regions. Table V-15, "1927 Shares of Tenant Land by Province," shows the variation in tenancy levels for paddy land and upland considered separately. Paddy tenancy was highest in North Jolla Province (78.8 percent), but was also extremely high in Kyonggi, South Chungchong and South Kyongsang (72.8, 71.5 and 76.0 percent respectively). Relative tenancy levels for upland and overall land were similar, and so from this data there is no reason to believe that tenancy conditions in Korea could have had much influence on regional productivity differences. Had owner-cultivation been dominant in Kyonggi we might have argued that individual incentives inspired adoption of techniques and exploitation of commercial potential. Landlord predominance in Kyonggi relative to their provinces, on the other hand, could have led us to conclude that owners were playing an important managerial role, coordinating information about both markets and technique. The data shed light on neither of these hypotheses.

There is one factor in the pattern of Korean tenancy, however, which although difficult to document may have had relevance for productivity because of its impact on tenant levels of work effort. Tenancy to Japanese landlords as opposed to Korean landlords was

---

was some variation. For one discussion of the levels and trends in tenancy see Kuro Komikawa, *Chosen Nogyo Hattatsu-shi, Hatten-ben,* cited above, pp. 62ff.

Table V-16.   Japanese and Korean Land for Redistribution, by Province

(%)

	Japanese			Korean			Total Land (hectares)
	Paddy	Upland	Total	Paddy	Upland	Total	
Kyonggi	18.0	11.1	15.3	82.0	88.9	84.7	217,462
Kangwon	8.6	4.9	6.0	91.4	95.1	94.0	30,989
N. Chungchong	23.0	16.4	19.5	77.0	83.6	80.5	45,576
S. Chungchong	30.7	21.5	27.9	69.3	78.5	72.1	110,979
N. Jolla	53.9	26.0	47.4	46.1	74.0	52.6	139,827
S. Jolla	52.0	43.7	50.3	48.0	56.3	49.7	145,965
N. Kyongsang	28.7	19.0	24.6	71.3	81.0	75.4	81,188
S. Kyongsang	38.9	30.1	36.6	61.1	69.9	63.4	91,142
Total*	36.2	20.8	30.9	63.8	79.2	69.1	864,127

Sources: Data for vested lands confiscated from the Japanese in 1945 are from Hiroshi Sakurai, *Kankoku Tochi Kaikaku no Saikenson* [A Reexamination of South Korean Land Reform], Tokyo: Asian Economic Research Center, 1976, (in Japanese), p. 65. Data for Land to be Purchased for Redistribution in 1949 are from Bank of Korea, *Saneop Chonggam* [Industrial Alamanac], Seoul, 1954, (in Korean), p. 523.

Note: The shares are of total land (vested and Korean) originally designated for redistribution as tenant land or land from holdings greater than 3 cheongbo.

*The total excludes consideration of Seoul and Jeju Island.

much higher in the rice belt areas of South Korea than it was elsewhere. The Japanese did not publish data separating Japanese and Korean landlord land ownership, but statistics compiled for the land reform after 1945 and data on land confiscated from Japanese after Korea's liberation give a clue to the regional differentiation in land ownership by nationality. Table V-16, "Japanese and Korean Land for Redistribution, by Province," presents these land reform data. The totals do not, of course, represent land actually distributed in the land reform. There was a considerable time lag between the early postwar surveys to determine land eligible for redistribution and the actual enactment of the land reform law in 1949 and its subsequent implementation during the years of the Korean War. During the intervening years, much of the tenant land was sold by owners to tenants under conditions much more favorable than those later afforded by the land reform law. For the purpose of determining Japanese-era ownership patterns, however, the original feasibility survey results are most relevant. If land ownership patterns in 1945 give some indication of what they were in the 1930's and before (and evidence suggests that that is likely),[19] Table V-16 shows that while Koreans owned more than 80 percent of tenant land in Kyonggi, they owned only half in the Jollas. Particularly striking is the very large share of tenant paddy owned by Japanese in North Jolla (53.9 percent) when compared to the same figure for Kyonggi (18.0 percent). The Japanese, then, had a very strong and perhaps dominating presence in the southern rice belt region which they did not have further north.

The contrast between Kyonggi and North Jolla is most significant, because analysis earlier in the century showed that it was North Jolla which had most clearly benefited from land improvement programs and chemical fertilizer consumption. Reports of tenancy disputes, however, which increased in number very rapidly in the 1930's, show a much heavier incidence in the Jollas and in particular North Jolla than in the other provinces.[20] The most important causes of contention concerned tenant rights to continue farming land and the level of rents, but their greater incidence in the Jollas and in particular North Jolla, when combined with the information on Japanese land ownership, provides reason to believe that tenant-landlord relations and farm working conditions may have been considerably more strained in the Jollas than in Kyonggi.

If this was so, there may have been considerable barriers to the effective use of new techniques and application of improved inputs in the southern regions. How significant for productivity might such barriers be? At most they could have counterbalanced the relative benefits of land improvement and fertilizer consumption, but it is not likely that better owner-tenant relations by themselves were responsible for the singular success of Kyonggi. Most telling in this respect is the comparison between Kyonggi and South Chungchong, which had been the most well-to-do province early in the century.

[19] As noted above, overall tenancy levels changed only gradually during the Japanese period. This implies nothing, of course, about the ownership mix hidden beneath the aggregate statistics, and it is possible that during the difficult 1930's the financial conditions of Korean landlords forced them to sell to Japanese in the rice belt. The hardships of the 1930's fell most heavily on the tenants, however, and it is not likely that regionally differentiated land sales took place on a scale large enough to invalidate conclusions drawn from Table V-16. For a source discussing trends in tenancy see note 18 above.

[20] See Choji Hishimoto, *Chosen Mai no Kenkyu,* cited above, pp. 129-132.

Although the same sources cited above show tenancy disputes were also higher in South Chungchong, the land ownership data do not suggest a particularly strong Japanese presence there. In sum, although the data on Japanese land ownership and tenancy disputes provide grounds for interesting speculation about the effects of farmer alienation on regional productivity, there is neither sufficient evidence nor substantiated causal reasoning for concluding that these relationships were the decisive ones. The last of our explanatory factors, then, has not given us an answer to the puzzle of regional agricultural development in South Korea under the Japanese.

The present chapter has shown that rural south Korea began from a state of rough regional equality and then developed in a way heavily favoring one region, the Northwest. The demographic, cropping and productivity patterns responsible are clear, but the causal forces are not. With larger farms, Kyonggi sharply increased its traditionally lower rice yields, and its endowment of upland was well-suited to the needs of a vegetable industry which by the 1930's dominated that in all other parts of the country. Equally as important, farm population growth in traditionally productive provinces was much greater than in the Northwest, eroding whatever increases in output from the land were realized. However, the usual sources of increased agricultural output, irrigation, new seeds and chemical fertilizers do not explain this unbalanced development, although increased application of chemical fertilizers may have been important. If we must point to a single responsible element in Kyonggi's relative success, it would be one several steps removed from the process of agricultural production itself. The complex influence of the city of Seoul on its rural hinterland must be considered the overriding force behind the gains made by Kyonggi farmers.

The importance of Seoul, with its large Japanese population,[21] for Kyonggi's vegetable market requires little explanation. Vegetables are perishable and somewhat bulky for their value, putting a premium on market proximity and convenient transport. The same reasoning cannot be used in the case of rice, however. Rice is easily stored and

---

[21] Out of a total population of 404,000 in 1935, Seoul had 113,000 Japanese residents, well over a quarter. These figures are originally from the 1935 Census of Population in Korea conducted by the Japanese. The percentage was even higher for Pusan, but the absolute numbers were smaller (180,000 total and 57,000 Japanese). The data are as published by the Chosun Government General, *Statistical Yearbook,* 1935, pp. 22-26.

is well-suited to bulk shipment by rail or by sea. Seoul's influence on Kyonggi's much-improved rice yields is therefore more difficult to understand, but may have involved better techniques and their more careful application, all spurred by the modern consumption and life-style influences of Korea's largest and traditionally most advanced urban center. Finally, the influence of Seoul's industrial labor market on demographic trends in rural Korea was probably very great, as farmers were drawn to the city for jobs and an improved standard of living. Available data make further generalizations about the mechanism of Seoul's influence on agriculture impossible. And yet it is a significant result of this chapter that such a relationship must have been important, regardless of its workings.

The Japanese colonial period in Korea, then, introduced these two significant characteristics of Korean agriculture's twentieth-century experience. Kyonggi assumed the position of superiority it was to hold and strengthen through the 1970's, and the influence of an urban center emerged as the only possible causal factor different enough for Kyonggi to offer an explanation for its success. Why was there not a comparable impact on South Kyongsang Province from the city of Pusan? In one important category South Kyongsang did mirror the experience of Kyonggi, for Table V-7 above shows that its farm household growth rate between 1927 and 1935 (0.5 percent) was the lowest of all provinces except Kyonggi (0.2 percent) and well below the average increase for South Korea as a whole (.12 percent). In the other major categories, however, South Kyongsang did not reflect the same influences as those felt by Kyonggi, but there are perhaps good reasons why.

In the first place, Pusan was not only smaller than Seoul, it was also not the center for wealth and luxury-living that Seoul was. Seoul was the administrative capital and the industrial center, in addition to being the traditional royal capital. Furthermore, Pusan was the only city of significance in South Kyongsang Province, while Seoul's urban area was supplemented by that of the rapidly growing port of Incheon. This difference and the very mountainous terrain of much of South Kyongsang cutting off the larger portion of the province's farmland from Pusan probably accounts for the low level of per-household vegetable production during this period. Data for the years after the Korean War, however, will show that as roads were built and as Pusan continued to increase in size, the same vegetable industry phenomenon makes its appearance.

When asking about South Kyongsang's poor success in increasing its rice yields, it is important to notice that rice yields in this heavily populated Southeastern region were already the highest in the country by 1923, and it is very likely that the potential for increase latent in Kyonggi's under-exploited condition at that time did not exist or had already been realized in South Kyongsang, where half of the same paddy soil was also required to supply nutrients for a winter barley crop (double-cropping in Kyonggi, we remember, was insignificant— roughly 2 percent—at that time). In other words, although both provinces had important urban centers, South Kyongsang's circum- stances were very different, and different in ways important for ex- plaining its lack of urban-influenced growth.

As interesting as the Japanese period might be for its unbalanced development and the apparent influence of Seoul, the pattern of regional growth before 1940 really only sets the stage for the more complex and better-documented trends following liberation and the Korean War. The patterns of regional evolution during South Korea's period of rapid industrial growth remained bracketed by the wealth of Kyonggi and the poverty of Kangwon, and within this overall frame- work shifts and trends among provinces in rural Korea of the later period will be shown to exhibit many of the same phenomenon which had begun to appear in the colonial era.

CHAPTER VI

REGIONAL FARM PRODUCT AND GROWTH IN SOUTH KOREA (1945-1975)

The decade of the 1940's found Korea caught up in a swirl of international and domestic events which altered her circumstances almost beyond recognition. These events of course had an impact on the regional pattern of the peninsula's farm economy, but by and large the trends established by the end of the colonial period emerged from the chaos to be confirmed and strengthened. If the focus of attention in the pre-war period was the performance of Kyonggi and the seeming importance of its relationship with Seoul, the most interesting regional phenomenon of the post-war period has been the very sudden and unmistakable eclipse of the rice-belt regions by the major upland provinces. The importance of access to urban industrial areas for explaining these developments is not as apparent as it is in the case of Kyonggi, but analysis in this and subsequent chapters will demonstrate that cities have spread their magic influence in these latter years as well.

As was the case in the pre-war period, it is relatively easy to describe the important components of the relative success of North Chungchong and North Kyongsang's farmers in the decades following the Korean War. Rice yields and the production of vegetables, fruits and special crops all make important contributions. By the same token, however, it is not easy to explain the success of these upland regions by relying on analysis of traditional inputs and sources of agricultural growth. Neither land, irrigation, fertilizer, pesticides, population movements nor even prices favor the regions which have done so well.

The present chapter will detail the major regional trends in rural South Korea during the thirty years following 1945. Although many parts of the presentation are descriptive, the statistics and analysis in the following passages make it quite clear that usual explanations for the expansion of agricultural production are not sufficient, and that we must look further if we hope to fully understand South Korea's farm experience.

The clearest picture of the period's major regional trends is provided by Table VI-1, "Per Capita Gross Farm Output by Province,1938-1974." The data in this table continue the pre-war series presented in Table V-1 of the preceding chapter, except that Seoul, Pusan and Jeju have been separated out. In addition, the figures in this table are per capita values rather than per household ones, and except for this difference the five-year averages for 1938 are very nearly compatible with the 1935-1940 three-year averages in Table V-1.[1] The per capita data confirm that Kyonggi had outpaced the other regions by the end of the colonial period, and that Kangwon had fallen behind the other regions. It is also interesting to note that except for Kangwon, the prewar differences between provinces had all but disappeared during the years following Japanese repatriation and American assumption of civil authority. In fact, during those five inter-war years, per person output in Kyonggi (29.1) and even in Seoul (30.0) was quite close to matching that in South Chungchong (25.9), North Jolla (28.3) and South Jolla (30.2)

After the Korean War, however, and in particular after 1955, Kyonggi resumed its previous position of dominance, and from that point on, interest shifts to the competition between rice-belt and upland provinces. If we take particular notice of North Chungchong and North Jolla as representative of their regions, we see that although the latter rice-belt province outproduced the former until the mid-1960's, after 1965 North Chungchong assumed an unmistakable

---

[1] The largest apparent discrepancy between the two tables results from comparing North Chungchong with South Jolla for the late 1930's. The larger gap with the per-household data results from South Jolla's somewhat smaller households. The average number of persons per farm household in each province for 1943 are given below:

KK	KW	NC	SC	NJ	SJ	NK	SK	Korea
5.9	6.0	5.9	5.6	5.6	5.5	5.9	5.7	5.8

These were calculated from data in Bank of Chosun, *Chosun Kyeongje Yeonbo,* Seoul, 1948, pp. 1-5.

Table VI-1.   Per Capita Gross Farm Output by Province, 1938-1974

(thousand 1970 won; 5-year averages)

	1938	1947ᵃ	1955	1960	1965	1970	1974ᵃ
Seoul	—	30.0	51.0	80.1	54.4	66.9	114.4
Pusan	—	—	—	—	34.3	41.7	69.5
Kyonggi	35.7	29.1	36.3	41.1	44.4	57.5	75.2
Kangwon	19.1	16.7	26.8	30.3	31.1	38.7	50.1
N. Chungchong	26.1	27.3	27.5	32.1	39.8	50.7	69.8
S. Chungchong	27.7	25.9	30.4	33.8	38.9	46.3	57.5
N. Jolla	26.7	28.3	35.2	37.3	42.2	48.6	55.2
S. Jolla	24.2	30.2	31.9	33.1	34.9	42.1	53.4
N. Kyongsang	22.9	26.8	33.5	33.8	40.7	50.0	64.4
S. Kyongsang	25.0	23.0	27.9	31.6	39.6	49.7	57.0
Jeju	—	24.7	24.6	30.0	39.2	45.4	102.4
South Korea	25.7	26.0	31.6	34.2	39.2	47.9	60.9

Sources: 1970 constant prices were applied to individual crop output levels to obtain gross output which was then divided by annual administrative figures for farm population before averaging. 1938 figures are from per-household data converted with the 1943 sizes of farm households by province. For details, see appendix B.

Note: ᵃ The figures above are five-year averages except for Seoul and Jeju for the year 1947 when they are 2- and 4-year averages respectively; for 1974 all figures are three-year averages.

and commanding lead. By 1974 North Chungchong's per capita product (69.8) was more than 25 percent greater than that of North Jolla (55.2).

It is also interesting to note the relative shifts in position among the rice-belt provinces themselves. Both South Chungchong and South Jolla lagged North Jolla throughout most of the post-Korean War years, but in the 1970's, South Chungchong, with its more northerly location, began to out-produce both of the others, while South Jolla remained in third place. If we are looking for signs of the influence of of Seoul, Incheon, Taejeon and other major cities, we have it here in the relative success of less isolated provinces. We can also find similar evidence in the experience of the second major upland province, North Kyongsang, with its very large industrial center at Taegu. Although its rise in per capita produce is not quite as dramatic as that for North Chungchong, the two provinces together clearly exhibit levels of output well above those of the rice belt provinces, although stll significantly lower than that for Kyonggi.

In sum, although their performance was anticipated by tendencies in the colonial period, the upland provinces of North Chungchong and North Kyongsang did in a sense "come from behind" to out-produce the traditionally mild and fertile rice-bowl regions. Why was this so? It might be suspected that the prices for the year chosen as a base, 1970, favored upland output, and there is some truth to the accusation, for vegetable prices in that year had been pushed to ridiculous heights by crop failures in 1969. However, even if current prices are used and deflated by a farm consumer index, the trends remain hardly affected. This can be seen for the period of rapid industrial growth by referring to Table VI-2, "Regional per Capita Gross Farm Income, 1959-1975." Although only a few representative provinces are covered, it is clear that the ascendency of the upland areas was more than a fluke coincidence of abnormal prices and skewed crop mix. The challenge for students of regional patterns in South Korea's farm growth, then, is to explain this clear-cut shift in provincial ranking.

The components of this post-war evolution are somewhat more difficult to disentangle than those for Kyonggi in the Japanese period. This is because no single crop or pair of crops can be assigned respon-sibility for the success of upland regions. There have, of course, been shifts in cropping patterns, and the expansion of vegetable production has been particularly noticeable. Table VI-3, "Shares of Rice, Barleys, Vegetables and Fruits in Total Agricultural Output, by Region, 1938-1974," shows the degree to which traditional food-grain crops have lost ground to crops grown more and more for the urban market. The country as a whole has experienced a large drop in the importance of rice (72 percent to 52 percent between 1947 and 1974), but this decline was much more rapid in the upland region re-presented by North Chungchong than in South Jolla, for example. Conversely, although there was an 11 percentage point increase in the share of national farm output commanded by vegetables and fruits, North Chungchong experienced an increase of over 20 percentage points between 1947 and 1974. In this sense, then, North Chungchong and the upland regions have been in the vanguard of nationwide movements to new cropping mixes, while the rice belt regions have lagged behind the prevailing trends.

The importance of these cropping share changes for provincial per capita farm product can be seen in Table VI-4, "Per Capita Output

Table VI-2.   Regional per Capita Gross Farm Income, 1959-1975

(thousand won in current prices deflated to
1970 prices by farm consumer index)

	Kyonggi	Kangwon	North Chungchong	South Jolla	South Kyongsang	South Korea
1959	30.6	21.2	22.5	23.2	21.8	21.5
1960	32.8	22.3	25.4	24.0	23.1	25.9
1961	37.3	25.5	32.2	32.1	31.4	32.9
1962	34.9	30.3	29.4	29.7	28.3	20.6
1963	50.5	32.1	38.9	34.9	34.2	39.5
1964	57.9	37.5	51.6	47.0	49.1	50.4
1965	38.9	29.2	38.9	36.3	46.7	40.0
1966	46.0	32.2	45.8	40.0	43.3	43.4
1967	46.4	36.8	42.8	30.1	42.1	39.7
1968	41.5	32.0	38.2	30.7	39.1	37.4
1969	55.2	35.1	41.9	41.3	48.9	45.7
1970	60.9	37.1	50.8	41.5	49.9	48.3
1971	63.4	42.5	55.6	49.2	55.2	52.9
1972	61.0	46.0	59.2	55.8	57.2	57.3
1973	70.9	51.9	64.3	55.3	54.2	58.9
1974	83.4	50.1	76.7	59.4	66.8	67.7
1975	95.2	66.3	83.7	75.6	74.3	79.6

Sources: See footnote to Table VI-1. For similar data covering all provinces see Appendix B.

Table VI-3.   Shares of Rice, Barleys, Vegetables and Fruits in Total Agricultural Output, by Region, 1938-1974

(%)

	Rice and Barleys				
	1938	1947	1960	1966	1974
Kyonggi	60.7	62.8	64.3	58.5	48.7
Kangwon	61.3	49.8	50.8	36.6	33.3
N. Chungchong	61.7	71.2	64.9	50.2	40.9
S. Jolla	72.1	77.8	67.3	62.2	59.9
S. Kyongsang	73.6	73.2	67.6	63.3	56.9
S. Korea	70.3	72.2	66.0	58.4	52.1

Table VI-3.   (Continued)

	Vegetables and Fruits				
	1938	1947	1960	1966	1974
Kyonggi	26.5	22.9	16.2	22.7	23.2
Kangwon	13.0	15.3	13.2	24.7	23.1
N. Chungchong	15.8	9.8	11.0	18.9	31.4
S. Jolla	7.7	6.2	9.1	11.8	13.0
S. Kyongsang	10.5	8.6	11.0	12.5	20.1
S. Korea	13.4	10.1	11.9	16.6	21.4

Source: See footnote, Table VI-1.

from Rice and Vegetables by Region, 1938-1975." The most striking numbers in this table are those comparing per capita rice output for North Chungchong and South Jolla in 1970 and 1975. For although food grain production as a share of total output fell more rapidly for the former than the latter, North Chungchong was nevertheless able to increase its per capita rice production to the point where it was greater than that in the "rice-bowl" province. As we will see, this is due in some degree to the great increase in barley output for southern regions during this period, but the fact remains that rice's (as opposed to foodgrains') share fell faster in an upland province while at the same time that region's per capita output surpassed the level in South Jolla. In other words, we have here again the same pattern of circumstances accompanying the disproportionately rapid growth of one region in relation to others which we observed for the 1930's. The surge in productivity spans foodgrain crops as well as cash crops for the urban market, and hence even if we concede that the influence of cities is relevant, we cannot pretend that the mechanism is merely better transport and hence less spoilage for perishable produce. Vegetables are certainly important, as Table VI-4 clearly shows, but attention must be paid to the overall increase in productivity observed.

Similar but more detailed results which at the same time are free from the bias of a single year's prices are given in Table VI-5, "Upland and Rice Belt per Capita Income from Selected Crops, 1959-1975," where crop output levels have been combined with current prices and then deflated by a farm consumer price index. Because the data are annual, temporal shifts in crop patterns are particularly

Table VI-4. Per Capita Output from Rice and Vegetables by Region, 1938-1975

(thousand won/capita, 1970 prices)

	Rice					
	1938	1947	1960	1966	1970	1975
Kyonggi	31.2	26.9	25.4	22.8	29.2	35.3
Kangwon	8.6	9.9	12.0	9.8	12.3	13.2
N. Chungchong	15.7	13.3	15.1	14.4	16.3	24.9
S. Jolla	18.8	16.6	17.0	15.7	18.5	21.1
S. Kyongsang	18.4	14.4	16.5	18.3	18.8	24.9
S. Korea	19.1	16.8	18.0	17.7	20.2	25.3
	Vegetables[a]					
	1938	1947	1960	1966	1970	1975
Kyonggi	12.3	9.2	6.7	9.1	14.4	16.1
Kangwon	1.4	3.4	3.8	6.9	7.5	9.8
N. Chungchong	4.4	2.3	3.1	8.0	12.7	19.6
S. Jolla	1.8	1.5	3.1	4.0	4.4	6.1
S. Kyongsang	2.3	1.8	3.3	5.1	9.8	10.0
S. Korea	3.6	2.6	3.4	5.8	7.8	10.1

Source : See footnote to Table VI-1.
Note: [a] Exclusive of fruits. For similar data covering other categories and all provinces see Appendix B.

clear, and special attention should be drawn to the very sudden increase in North Chungchong's vegetable output between 1969 and 1970. We will see in a later chapter that this surprising change corresponds precisely with the completion of high-speed highway construction connecting parts of the province with Seoul.

The same table is also useful for showing the degree to which South Jolla at the same time expanded its output and income from barley. This trend is a direct response to the introduction of subsidized prices for food grain crops, and the resulting increased barley output in South Jolla in fact more than compensates for the upland surge in vegetables, the effective prices of which, as we saw in Chapter III, were actually falling. We have, then, government-sponsored expansion in barley output on the one hand more than matching market-generated increases in vegetable production on the other. With the

Table VI-5.   *Upland and Rice Belt per Capita Income from Selected Crops, 1959-1975*
(thousand won in current prices deflated to 1970 values by indexes of prices paid by farmers)

	Barley		Vegetables		Rice (current prices)		Rice (leading prices)	
	North Chungchong (i)	South Jolla (ii)	North Chungchong (iii)	South Jolla (iv)	North Chungchong (v)	South Jolla (vi)	North Chungchong (vii)	South Jolla (viii)
1959	4.1	3.6	1.5	1.5	10.6	12.2	12.1	13.4
1960	4.6	4.0	2.0	1.7	12.0	12.4	13.9	14.0
1961	6.5	6.2	1.7	2.1	14.7	17.5	15.1	17.5
1962	5.3	6.4	2.7	1.4	10.0	14.8	13.4	18.5
1963	3.8	2.1	4.7	1.6	18.0	22.2	19.0	24.0
1964	10.3	8.9	3.8	2.9	19.5	24.5	18.1	21.8
1965	6.0	6.8	4.7	3.4	13.8	16.4	13.3	15.8
1966	8.2	6.5	7.2	4.3	14.8	17.9	14.5	17.6
1967	7.4	7.1	6.3	2.7	13.1	11.3	13.0	11.2
1968	5.4	8.0	6.8	2.9	11.8	9.0	12.7	9.7
1969	6.4	8.1	6.5	2.9	14.9	19.1	14.7	18.8
1970	5.8	8.6	12.4	3.9	15.7	17.0	16.7	18.1
1971	6.1	10.8	11.4	4.2	20.5	21.8	22.0	23.4
1972	6.7	14.4	10.5	3.8	20.3	24.1	19.8	23.5
1973	5.9	12.2	10.3	3.7	23.5	23.9	24.7	25.1
1974	4.7	11.5	12.7	4.4	31.0	26.2	31.7	26.9
1975	5.6	14.1	13.7	6.0	38.0	35.0	36.9	34.0

Sources: See footnotes to Table VI-1 and the text. "Leading prices" in the two right-hand columns refers to the use of both rice prices and farm consumer price deflators which are weighted sums, 40 percent of the current year's price and 60 percent of the following year's price. See the text for a discussion of this reasoning.

contribution of vegetables in this way "neutralized," we must search further in order to understand the output composition responsible for upland success.

Before commenting further on that composition, however, it is valuable to notice that Table VI-5 presents two separate series for the per capita output of rice valued at current prices. This is because the use of an average rice price for the output of the same calendar year introduces a certain degree of error, given the timing of harvest and marketing for South Korean rice. The harvest is in October and November, and as much as 60 percent of a year's rice harvest is actually sold by the farmer in the following year,[2] by which time prices have had a chance to adjust in light of the harvest's size. The last four columns in Table VI-5 allow comparison between rice production valued at harvest-year prices on the one hand, and a weighted average of prices in the harvest year and in the following year on the other.

The significance of using the more accurate second pricing method can be seen by looking at the results for two years: 1962, which was a bad harvest in all regions, and 1967, which was a bad year in the South but not in the North. As a result of the poor harvest in 1962, rice prices rose markedly in 1963, and when 60 percent of output is valued at the 1963 average price it actually brought, South Jolla's farmers received more income from rice when compared to 1961 rather than less, as the data using harvest year prices would suggest. The percentage rise in the effective rice price was greater than the percentage fall in South Jolla's rice output. The 1967 failure of the southern rice crops because of drought reveals a different phenomenon. Because the poor harvest was more localized, it did not have as great an effect on total national output nor on rice prices in the following year. In fact, the terms of trade for rice deteriorated in 1968 and the regional impact of the bad weather was made even more severe. The adverse movement in the purchasing power of rice was easier to bear in the northern regions, since the harvest there was respectable. In other words, the inclusion in our analysis of the actual price movements associated with poor harvests shows that the impact of such a poor harvest is a much more serious threat to farm income if it is localized than if it is widespread enough to increase rice prices

---

[2] See Ministry of Agriculture and Fisheries, *Report on the Farm Household Economy Survey*, Seoul, any year. The appendices give average sales per month for major crops.

throughout the country.

Correcting the per capita output figures in this way, however, does not alter the conclusion first drawn from the fixed price results: in addition to excelling in vegetable production as expected, rice production in upland regions also outperformed at least one rice belt province. Since increased barley production in the South kept in step with expanding vegetable output, the shifting rice productivity is highly visible as one contribution to the forces responsible for upland ascendency. Similar contributions are made by growth in the output of fruits, special crops, and silkworm which were not fully paralleled by South Jolla's nevertheless significant increases in the production of potatoes and pulses. In other words, North Chungchong's performance (and that of North Kyongsang as well) has been one of across-the-board growth unmatched by the combination of output increases in the paddy belt areas. We shall see in the following paragraphs that there is no clear-cut answer from traditional sources as to why this should be so.

We discovered in our study of regional product per household in the colonial period that differences from region to region in farm population density were very important for understanding both the rough equality at the beginning of the century and the reasons for regional divergence in the 1930's. Furthermore, studies by Nicholls and Tang of American farm communities showed clearly that if a region's industrial economy is stagnant or growing only very slowly, then population movements can in effect keep up with the changing ability of the land to provide a basic livelihood, leveling potential for poverty through migration.[3] These same studies showed, however, that if changes in the economy are rapid enough, population adjustments will be unable to keep up, and inequality between farm communities will ensue.

Rural South Korea in the period under study provides a clear confirmation of this phenomenon for a developing country. For although there is great disparity in farm population density for South Korea, the disparity has been gradually reduced as the movement of persons has attempted to compensate for much more rapid change in the value-productivity of Kyonggi's soil. We will notice in particular that the upland provinces have been only slightly more successful than the rice belt ones in increasing per capita output by reducing the size of

---

[3] See Nicholls, *op. cit.*

Table VI-6. Farm Population as a Share of Total Population by Province, 1949-1975

	1949	1955	1960	1966	1970	1975
Kyonggi[a]	71.7	60.5	57.2	46.3	43.5	33.3
Kangwon	60.2	55.6	49.2	47.8	46.6	39.1
N. Chungchong	83.8	75.9	73.9	71.7	69.1	58.3
S. Chungchong	83.5	75.5	73.6	73.5	63.8	53.7
N. Jolla	80.4	69.0	68.4	66.6	68.2	59.2
S. Jolla	81.4	74.0	73.5	71.1	66.9	54.1
N. Kyongsang	76.9	68.5	67.0	63.3	55.2	44.6
S. Kyongsang[a]	83.1	74.1	69.0	68.0	64.1	50.0
Jeju	84.8	81.1	82.2	75.7	67.9	56.1
S. Korea[b]	71.1	61.8	58.3	54.1	45.9	35.3

Sources: Total population figures for 1949-1966 are from Kwon, Tai-hwan, *"Population Change and Its Components in Korea, 1925-66,"* Ph. D. Dissertation, Australia National University, 1972, for 1970 and 1975 total population figures are from Economic Planning Board, *Korea Statistical Yearbook,* years 1973 and 1976. The 1970 data are based on the 1970 population census. Farm population data for 1949-66 and 1975 are taken from the Ministry of Agriculture and Fisheries, *Yearbook of Agriculture and Forestry Statistics,* various years. The farm population data for 1970 were taken from the 1970 Agricultural census.

Notes: [a] Data for Kyonggi Province and South Kyongsang Province do not include the populations of Seoul and Pusan nor the farm populations of those cities.

[b] The total includes all urban population, so it is not merely a weighted sum of the provincial figures shown.

the total farm population.

Statistics make it clear that the years since World War II have seen a marked decline in South Korea's farm population, both in a relative sense and, since 1968, in an absolute sense. Table VI-6, "Farm Population as a Share of Total Population by Province, 1949-1975," shows that the share of the country's total population in farming has declined from 71 percent to 35 percent in the course of a quarter century. With the exception of Kyonggi, however, it is difficult to discern any meaningful regional pattern from these data, in large measure because the presence of a large city on one side or another of a provincial border can wash out the relevant province's representative share. A somewhat better picture is obtained from county data, and Map VI-1, "Counties with Greater than 80 Percent of Total Population in Agriculture, 1970," shows that the South and much of

North Chungchong province have more than 80 percent of all persons in farming, but that in Kyonggi and around the city of Pusan more people do no direct farming and hence rely on the market to a much greater extent for their food.

A better idea of a farmer's demands from his land by region is given in Table VI-7, "Farm Population Density, 1933-1975," and in Map VI-2, "Farm Population per Hectare of Arable Land, 1970." As we would expect from our knowledge of densities in the colonial period, Kyonggi's density is much below that of other provinces. The same is true to a lesser degree for North Chungchong, though not for North Kyongsang, and when it is noticed that the rate of decline in density was greater for North Chungchong than for all other provinces except South Kyongsang, we have found yet another contributing though hardly compelling factor in the determination of overall per capita product. In fact, the only significant conclusion to be drawn from these data is that all the provinces of South Korea are reducing their farm population in comparison to that for Kyonggi, and that if this process were allowed to catch up with changes in productivity, the regional differences in per capita product might in fact disappear. That such a reduction is occurring can be seen clearly in Table VI-8, "Index Numbers of Farm Population Density, 1933-1975." Where South Kyongsang's density was 81 percent greater than Kyonggi's in 1949, by 1975 it was only 41 percent greater. We can conclude, therefore, that Kyonggi is maintaining its superiority in spite of this shift in farm population density and its equalizing influence.

Our examination of farm population movements in South Korea, then, provides us with some basis both for understanding North Chungchong's experience and for appreciating the strength of productivity gains in Kyonggi responsible for its continued dominance. To what degree can we say that land endowment and the treatment of land with chemical preparations has also enhanced the fortunes of the increasingly more affluent areas? On the surface of it there seems to be little if any at all significant differentiation in this respect. We have seen in Chapter IV (The Regions of South Korea) that the South and rice belt provinces have the superior endowment in paddy and double-cropped paddy, favored as they are by geography and climate. Table VI-9, "Fertilizer and Irrigation," and Maps VI-3 and VI-4, "Fertilizer Application per Hectare of Arable Land, 1970" and "Irrigated Paddy as a Share of Total Paddy" confirm that the

Map VI-1. Counties with Greater than 80 Percent of Total Population in Agriculture, 1970

(Higher numbers represent counties with correspondingly higher shares of population in agriculture.)

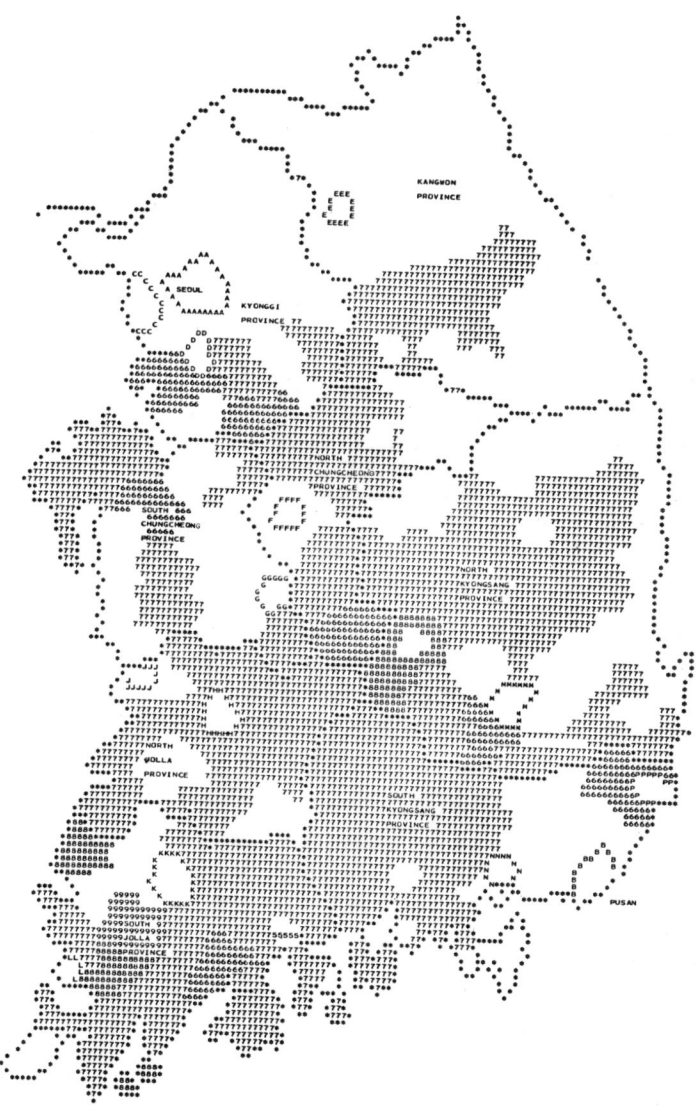

Table VI-7.   Farm Population Density, 1933-1975

(persons/km² of cultivated land)

	1933	1949	1955	1965	1975	Annual Percent Change, 1965-75
Kyonggi	333	509	499	550	449	− 2.0
Kangwon	385	559	563	590	470	− 2.2
N. Chungchong	492	665	646	672	504	− 2.8
S. Chungchong	471	728	684	707	542	− 2.6
N. Jolla	537	729	631	699	582	− 1.8
S. Jolla	498	805	722	758	596	− 2.4
N. Kyongsang	515	721	678	732	567	− 2.5
S. Kyongsang	565	919	837	857	632	− 3.0
Jeju	—	585	561	544	468	− 1.5
S. Korea	469	702	662	701	547	− 2.5

Sources: 1933 data are computed from statistics in Chosen Sotoku-fu [Chosun Government General], *Showa 8-Nen Nogyo Tokei Hyo* [1933 Agricultural Statistical Tables], Seoul, 1937, (in Japanese), pp. 7, 9. Other years: calculated from statistics in various years of the Ministry of Agriculture and Forestry, *Yearbook of Agriculture and Forestry Statistics,* Seoul.

Table VI-8.   Index Numbers of Farm Population Density, 1933-1975

(Kyonggi = 100)

	1933	1949	1955	1965	1975
Kyonggi	100	100	100	100	100
Kangwon	116	110	113	107	105
N. Chungchong	148	131	129	122	112
S. Chungchong	141	143	137	129	121
N. Jolla	161	143	126	127	130
S. Jolla	150	158	145	138	133
N. Kyongsang	155	142	136	133	126
S. Kyongsang	170	181	168	156	141
Jeju	—	115	112	99	104
S. Korea	141	138	133	127	122

Source: Calculated from Table VI-7.

Map VI-2.   *Farm Population per Hectare of Arable Land, 1970*
(Higher numbers represent counties with correspondingly higher densities,
and blank areas are counties with the lowest densities.)

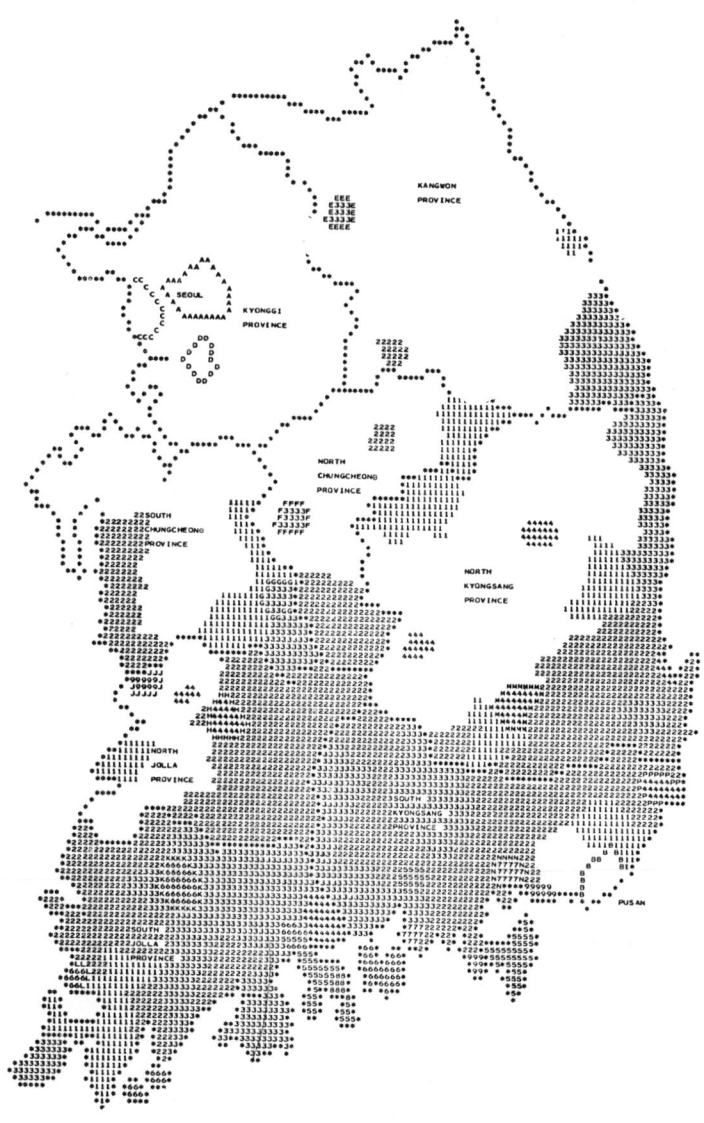

Table VI-9.   Fertilizer and Irrigation

	Chemical Nutrient Consumption (kg/ha.)			Securely Irrigated Paddy as a Fraction of Total Paddy (%)		
	1955	1965	1975	1955	1965	1975
Kyonggi	56	132	321	33.9	32.6	43.6
Kangwon	43	133	330	48.4	53.1	57.5
N. Chungchong	62	160	320	29.6	34.5	47.7
S. Chungchong	58	163	351	29.3	31.7	40.1
N. Jolla	62	191	438	22.0	27.3	32.1
S. Jolla	64	199	418	30.9	37.7	49.6
N. Kyongsang	68	167	363	33.7	38.8	54.5
S. Kyongsang	73	228	405	30.5	35.9	48.2
Jeju	56	171	438	39.2	39.8	71.1
South Korea	62	174	374	30.8	35.1	45.9

Sources: Calculated from chemical nutrient (nitrogen, phosphorous and potassium) consumption data, irrigated paddy data and total cultivated land from Ministry of Agriculture and Forestry, *Statistical Yearbook of Agriculture and Forestry,* relevant years. For more detailed data, see Appendix B.

upland regions are not especially favored in these respects either. Those provinces consuming more fertilizer are precisely the rice belt and southern ones, while the fact that Kangwon enjoys the best irrigation of all the mainland provinces casts some doubt on the importance of this factor for explaining overall farm output. Finally, we are prompted to reach similar conclusions regarding the possibility that regional differences in pesticide application have had a telling influence on regionally differentiated output when we notice in Table VI-10, "Pesticide Consumption per Hectare of Cultivated Land," that the rice belt and southern areas have benefitted most. This fact is illuminated clearly in Map. VI-5.

In response to the evidence presented in the preceding paragraph some might say that when consideration is made of the greater intensity with which land is used in the heavily double-cropped South, levels of fertilizer and pesticide application may not in fact favor those regions. We are, after all, ultimately interested in the degree to which the distribution of these inputs has made one region more able

Map VI-3. *Fertilizer Application per Hectare of Arable Land, 1970*
(Higher numbers represent counties with correspondingly higher
intensity of fertilizer application.)

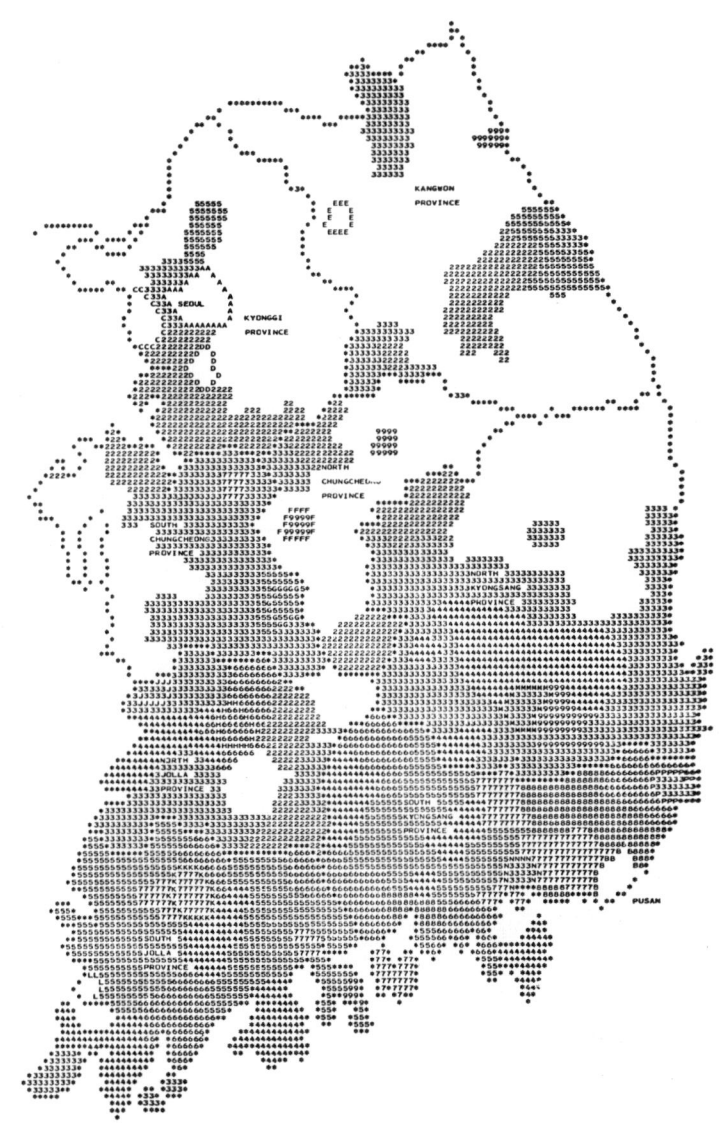

Map VI-4.   Irrigated Paddy as a Share of Total Paddy
(Higher numbers represent counties with a correspondingly greater
share of irrigated paddy.)

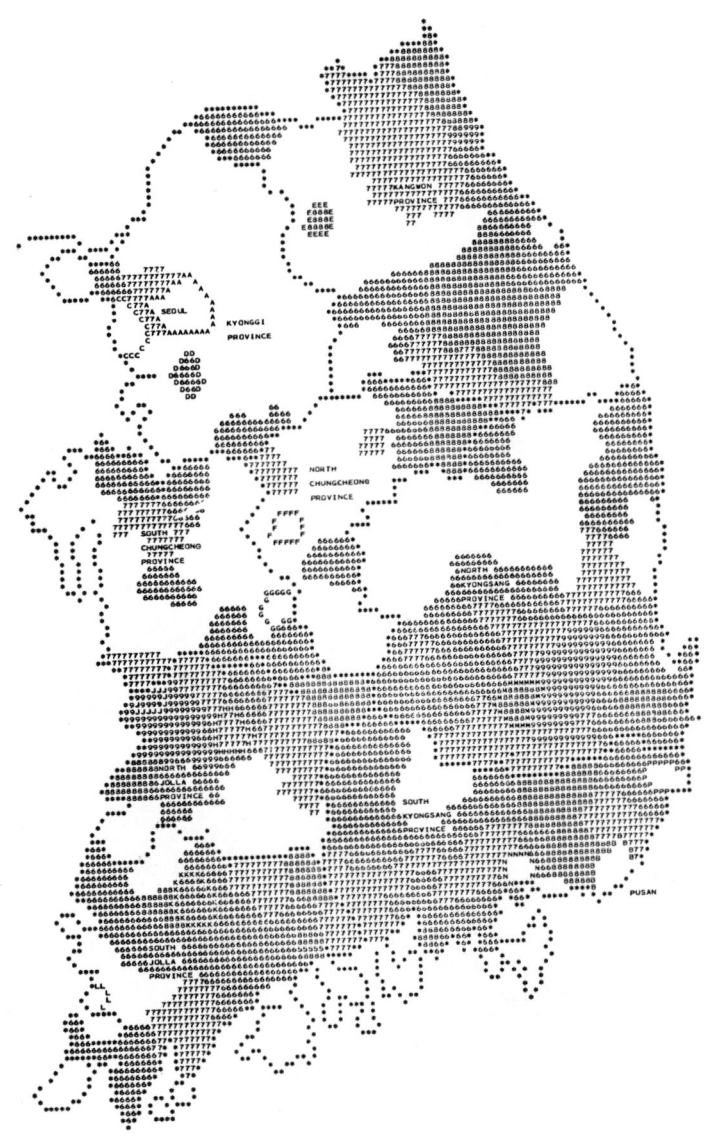

Map VI-5. *Pesticide Spray Equipment per Hectare of Cultivated Land, 1970*
(Higher numbers represent counties with a correspondingly heavier use of
spray equipment).

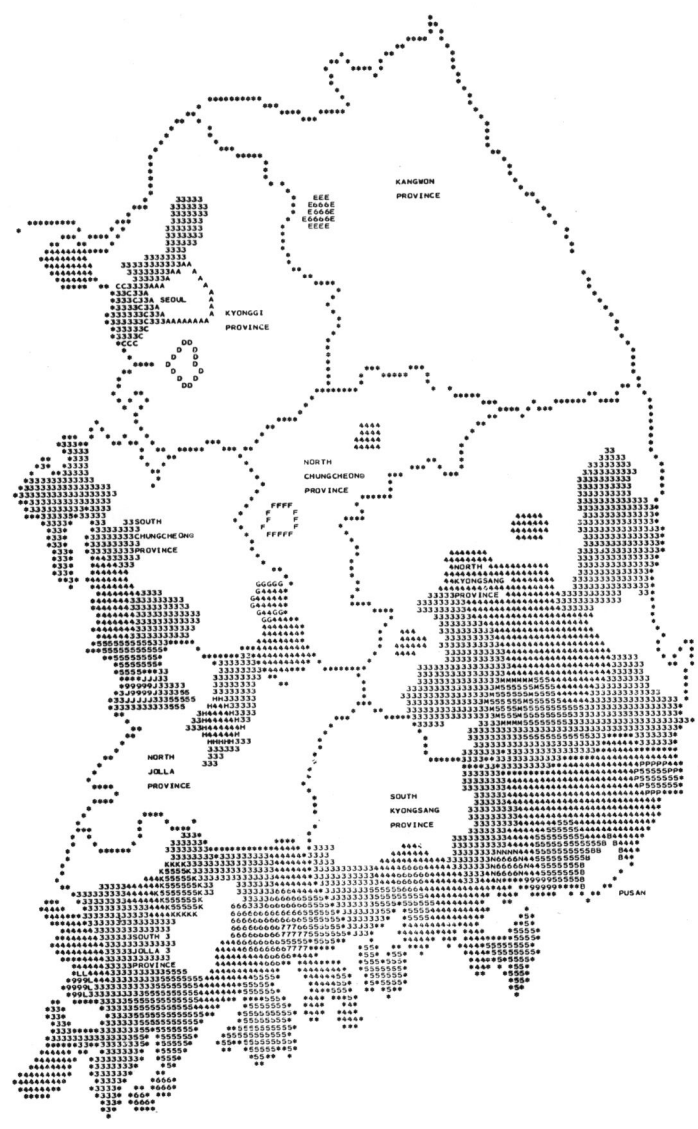

Table VI-10.   Pesticide Consumption per Hectare of Cultivated Land

(kg/ha.)

	1955	1965	1975
Kyonggi	1.6	4.7	32.8
Kangwon	0.2	2.2	10.5
North Chungchong	1.0	5.2	25.1
South Chungchong	0.8	5.9	29.2
North Jolla	1.6	6.0	30.6
South Jolla	2.3	6.1	29.3
North Kyongsang	5.0	8.3	29.6
South Kyongsang	1.4	5.0	28.7
Jeju	0.0	0.6	16.0
South Korea	2.0	5.6	28.0

Sources: Provincial pesticide data in Ministry of Agriculture and Forestry, *Statistical Yearbook of Agriculture and Forestry Statistics,* relevant years. For more detailed data, see Appendix B.

than another to support a member of its farm community. A check of the regional differences in per capita (as opposed to per hectare) applications of chemicals reveals that even by this measure both fertilizers and pesticides are ministered more heavily in the rice belt than in upland provinces.[4]

In spite of these findings, it is perhaps wrong to say that natural endowments have given no advantage to the upland areas. If paddy land is planted to rice, it produces an extremely high value-yield per hectare-year when compared to that for the usual uses of up-land. In fact, in the immediate vicinity of Seoul paddy is still invariably planted to rice rather than a series of summer vegetable crops. When

---

[4] The calculations were made for 1975 from Tables VI-7, VI-9, and VI-10. The results, in kg/person, are given here:

Fertilizer:

KK	KW	NC	SC	NJ	SJ	NK	SK	Korea
72	70	64	65	75	70	64	64	68

Pesticides:

KK	KW	NC	SC	NJ	SJ	NK	SK	Korea
7.3	2.2	5.0	5.4	5.3	4.9	5.2	4.5	5.1

asked directly the reason for doing so, farmers always said that they got more money that way. A new opportunity to grow and sell vegetables is not such a godsend if he is already planting his land in rice. If, however, his land was not good enough to use as paddy, a market for vegetables represents a suddenly much more profitable use of existing fields. In this sense, upland areas, less well endowed in paddy, are really better endowed for taking advantage of opportunities offered by access to the city. This is of course only one more variation on the theme that centuries of equilibrating adjustments settled South Korean farm population in densities which reflected the productivity of the soil and that developments in the twentieth century have upset that balance at a pace more rapid than can be dealt with by compensating population movements. This point made in reference to vegetables, however, still does not respond to our need for an explanation of upland increases in the production of rice and other crops not so dependent on ready access to city markets. This other more general secret of upland success will in fact continue to elude us throughout the rest of this study.

Our final attempt at explaining post-war regional production patterns through traditional arguments will examine the regional significance of price differences for different farm products. If shifts in prices suddenly favored farmers in the upland areas, we would have good reason to believe that the response to these prices could be elastic enough to result in considerable gains in production. The paragraphs which follow will show, however, that far from encouraging output in the upland areas, price changes rather acted in the favor of rice belt provinces.

As we noted in Chapter III, if the prices a farmer receives for his products go up faster than the prices for goods he buys, his terms of trade are improving and he is both better off and more able to turn his own efforts into useful goods and services. Terms of trade are therefore the ratio of indexes for prices received and prices paid, where the indexes are compiled as a weighted sum, the weights usually being each good's share in total value using prices from one or several periods. For the purpose of regional study in this chapter, a "purchasing power ratio"[5] rather than the strict terms of trade will be used. Purchasing power ratios are obtained by dividing output valued

---

[5] See Table III-5, "Purchasing Power Ratios by Major Commodity Groups, 1959-1975."

Table VI-11.   Price and Purchasing Power Indexes, 1959-1975

(1970 = 100)

	Farm Consumer Price Index (i)	N.A.C.A. Index of Prices Received (ii)	N.A.C.A. Farm Terms of Trade (iii)	Purchasing Power Ratios (iv)
1959	24.8	17.4	70.2	72.3
1960	26.6	20.9	78.6	81.1
1961	28.7	24.6	85.7	88.4
1962	31.8	27.1	85.2	90.9
1963	35.3	40.1	113.6	120.2
1964	44.8	50.2	112.1	126.7
1965	51.8	52.2	100.8	104.5
1966	58.1	55.4	95.4	98.4
1967	65.8	63.5	96.5	96.3
1968	78.8	74.3	94.3	88.8
1969	86.8	84.8	97.7	94.6
1970	100.0	100.0	100.0	100.0
1971	114.4	121.4	106.1	104.7
1972	130.5	147.9	113.3	111.8
1973	143.1	164.2	114.7	110.8
1974	192.5	215.6	112.0	112.6
1975	237.6	267.6	112.6	114.6

Sources: Columns (i), (ii) and (iii): Nong-eop Hyeop-dong Cho-hap Chung-ang Hwei Chosa-bu [National Agricultural Cooperative Association (N.A.C.A.) Research Division], *Nong-chon Mul-ka Chong-ram* (1959. 1-1974. 6) [An Overall View of Farm Village Prices (1959. 1-1974. 6)], Seoul, 1974, (mimeographed, in Korean), p. 4 and other unpublished N.A.C.A. statistics.

in fixed base year prices into output valued at current prices deflated to that same base year by a farm consumer price index. The resulting index is not the terms of trade in the strict sense, because the methodology in effect amounts to the use of pseudo-weights which do not sum to unity as true weights should. The resulting purchasing power ratios are nevertheless a good indication of the movement of crop prices in relation to the farm cost of living. In fact, Table VI-11, "Price and Purchasing-Power Indexes," shows that the ratios for all of South Korea move in close step with the farm terms of trade

Table VI-12. Regional Purchasing Power Ratios

(1970 = 100)

	Kyonggi	Kangwon	North Chungchong	South Jolla	South Kyongsang
1959	72.5	72.2	72.5	72.1	71.9
1960	80.3	80.3	80.8	80.4	81.9
1961	85.8	85.2	89.4	87.6	88.7
1962	85.5	86.3	92.7	91.3	91.8
1963	118.9	116.2	123.0	119.4	118.4
1964	120.7	118.5	129.6	126.3	130.6
1965	110.1	103.9	105.7	103.2	106.9
1966	97.5	100.4	101.8	96.1	97.7
1967	94.2	101.8	93.7	96.3	97.6
1968	86.2	90.2	80.7	91.7	91.7
1969	93.1	92.9	89.0	96.5	94.2
1970	100.0	100.0	100.0	100.0	100.0
1971	100.0	101.5	101.6	108.0	105.7
1972	104.4	107.9	108.3	117.6	111.1
1973	105.2	112.4	105.3	116.1	110.7
1974	108.5	110.9	106.8	117.0	113.8
1975	112.7	112.0	109.0	121.9	116.4

Sources: See text (page 172) and footnotes to Table III-5. For similar data on all provinces see Appendix C.

calculated by the South Korean National Association of Agricultures Cooperatives.

We have also seen in Chapter III that the purchasing power ratios differed for different crops in the years 1959-1975, and that in particular the terms of trade for vegetables seem to have fared much worse than those for rice and barleys. The combined effect of these and other different terms of trade for individual crops and the regional mixes in farm output together provide us with distinct purchasing power ratios for each province. These are presented in Table VI-12, "Regional Purchasing Power Ratios," and the sharpest contrast in regional terms of trade is between those for North Chungchong Province and South Jolla. After a long period of relatively similar changes for the two provinces in the 1960's, the terms of trade for South Jolla improved dramatically after 1970. At the same time, those for the upland region also improved, but at a slower rate. When these figures are compared to the average national ratio terms of trade in the fourth column of Table IV, it becomes clear that farmers in the three northern provinces shown here all had slower increases in the purchasing power of their output than did farmers in the south. When we further remember that the improved terms of trade for rice and barley are largely the result of government policies subsidizing their sale, it becomes clear that the upland areas have in fact been the objects of discrimination from policies meant to encourage Korean self-sufficiency in foodgrains. We can therefore say that differential price movements can have had little to do with the eclipse of the rice belt by North Chungchong and North Kyongsang.

The different terms of trade faced by farmers in different regions introduced above resulted from assuming that the actual prices they received were the same from region to region, and that only the shares of different crops in total output influenced regional variation. It is also interesting to see if the actual prices themselves are different from one region to the next. Data on regional differences in actual prices received by farmers are difficult to obtain, but Table VI-13, "1970 Rice Prices by Rural Markets" shows that if we take the least isolated of rural markets for which data are available, there does exist some difference in the actual price of rice received by farmers. The price received by farmers in Kyonggi province, which is close to Seoul, is almost 5 percent higher than that received in most other provinces. This is not surprising, given the tremendous importance of

Table VI-13.   1970 Rice Prices by Rural Markets

(won/100 liters of middle grade rice)

Province	Market Location	Price
Kyonggi	(Hwasong)	6,394
N. Chungchong	(Chongwon)	5,750
S. Chungchong	(Nonsan)	6,067
N. Jolla	(Okku)	5,710
S. Jolla	(Kwangsan)	5,882
N. Kyongsang	(Yongchon)	5,855
S. Kyongsang	(Changwon)	5,703

Source: Nong-eop Hyeop-dong Cho-hap Chung-ang Hwei Chosa-bu, *Nong-chon Mul-ka Chong-ram (1959-1970),* Seoul: National Agricultural Cooperative Association, 1971, (in Korean), p. 102.

Seoul in the Korean rice market. But it is interesting to note that North Chungchong Province does not seem to have benefitted from rice prices higher than those of the rice bowl provinces, South Chungchong and North and South Jolla. We can therefore conclude, that the terms of trade disadvantage of the upland dry field provinces induced by their greater dependence on non-food grain crops is not mitigated by regional variation in actual prices received for rice, which is still the most important crop in their output structure.

In South Korea's years of most vigorous growth and industrial expansion her rural economy has also become caught up in the throes of structural change and regional reorientation. The inherited extremes in status of Kyonggi and Kangwon continued to dominate the overall pattern of regional inequality in the years following World War II, but within the confines of these two extremes the remaining provinces seemed governed by forces bent on giving dry-field upland areas a distinctly superior productive position. Farm technology may have played some role, since upland areas benefitted from efforts to improve tobacco and red pepper techniques as well as from the early introduction of vynal green-housing. Subsequent statistical analysis, however, will show that the significance of vynal houses is not a major explanation for the production differences observed. Examination of the cropping changes and output growth responsible for the regional realignment reveals that although vegetables are clearly very important, there is no causal framework explaining the ascendency of North Chungchong and Kyongsang Provinces.

Just as our examination of the Japanese colonial years left us with the feeling that cities must somehow be important, so in South Korea's decades of rapid industrial growth are we drawn to the conclusion that Seoul and other modernizing centers must in some way be responsible for much of the regional divergence in South Korea's twentieth century. The continued power of Kyonggi's output expansion and the experience of satellite provinces rich in fields ready for high-yield use make it almost impossible to deny the significance of contact with cities for sustained South Korean farm growth. The chapters which follow will give a much more detailed analysis of isolation from cities and its importance for farm output.

CHAPTER VII

TRANSPORTATION AND RURAL ISOLATION

Throughout the twentieth century, under both Japanese colonial rule and Korean self government, the development of the peninsula's rural economy has strongly reflected the presence of urban industrial centers. Nevertheless, it is reasonable to believe that the mere presence of such centers is in itself not decisive, for rural areas must have channels of access to existing cities if they are to benefit from urban stimulation. For South Korea, appreciation of this fact requires some knowledge of the transportation network as it has grown in the century between 1876 and 1975, railroads on the one hand and, more importantly, highways and paved highways on the other.

Although it is difficult to demonstrate the regional impact of transportation routes on rural development, for South Korea it is less difficult for highways and paved highways than it is for railroads. The present chapter will briefly review the construction of railroads and highways under the Japanese before concentrating on the incidence of highway paving after 1945 and a measure of individual county isolation from industrial centers in 1970. The measurement of overall county isolation represents a serious effort at quantifying the phenomenon of isolation and potential for urban influence, and the results are both significant and central to the confirmation of this study's most general conclusions.

As we have already seen in Chapter II, transportation and communication networks at the end of Korea's Yi Dynasty were in desperate condition. By contemporary reports[1] there were few roads

---

[1] See Bank of Chosun, *Economic History of Chosun,* Seoul, 1920, pp. 101ff.

Table VII-1. *Korean Railroad Statistics, 1900-1943*

	Track (kilometers)	Passengers (million persons)	Freight (millions tons)	Average Distance per ton (km.)	Rice Shipped (1000 tons)	Fertilizer Shipped (1000 tons)
1900	29	—	—	—	—	—
1907	481	2.6	.4	102	56	.6
1909	980	1.9	.7	130	72	2.0
1911	—	2.4	1.1	137	77	3.6
1913	—	5.0	1.4	143	161	8.9
1915	1,621	5.0	1.7	175	263	12.1
1917	1,757	7.1	2.5	259	276	27.5
1919	1,856	12.2	3.6	250	348	42.6
1921	1,875	13.8	3.3	213	435	25.8
1923	1,914	16.8	4.2	226	608	23.3
1925	2,107	18.2	4.3	219	712	47.3
1927	2,345	20.1	5.6	208	900	90.2
1929	2,752	23.2	6.2	210	822	140.8
1931	3,009	19.7	6.0	215	1,033	157.6
1933	2,935	22.2	7.3	205	927	258.1
1935	3,390	29.3	8.7	206	726	489.3
1937	3,737	35.9	11.4	235	863	621.8
1939	4,090	59.7	15.9	238	438	710.7
1941	4,463	90.0	20.8	239	1,021	741.7
1943	4,568	128.5	23.9	350	—	—

Sources: Calculated from Chosun Government General, *Statistical Yearbook* for 1915, 1927, 1935, 1943. The last year, 1943, was published in 1948 by the South Chosun Provisional Government.

much better than paths, and certainly no railroads or highways as they are commonly conceived. The Japanese therefore placed very large and early emphasis on the development of a comprehensive railroad and highway system. Most of the principal trunk rail lines were completed before annexation in 1910, and the heaviest period of road construction and improvement came in the subsequent decade at the same time as the land survey.

Table VII-1, "Korean Railroad Statistics, 1900-1943," shows that the extension of track beds continued at a steady pace throughout the Japanese period, and that both passengers and freight increased through to the mid-1930's when the beginnings of Japan's war economy footing began to be felt in the all-important railroads. The time of most interesting rural growth in the Japanese period was from the late 1920's to mid-1930's, and yet there seems little in the railroad statistics to single out those years. An exception in this respect is the series of data for fertilizer freight hauled by rail. The rapid increase in fertilizer deliveries after 1925 confirms conclusions drawn from other data on commercial fertilizer consumption. The increase in rural productivity during those years is most likely a direct result of increased chemical fertilizer application, but railroad data on fertilizer deliveries by line and by station which might shed light on fertilizer's regional impact are not available.

What was the importance of this railroad network for bringing rural Korea closer to cities? That is difficult to say. Rail lines linked major cities and had numerous intermediate stops (for example, on the Seoul-Pusan trunk line there were 57 stops, averaging one every eight kilometers).[2] For South Korea the principal route went South from Seoul and then split at Taejeon, in South Chungchong Province, with one branch going East and then South to Pusan and the other branch heading Southwest from Taejeon for the port of Mokpo in South Jolla. There were numerous government and private branch lines which increased in number through the years and which seemed to be more prevalent in the southern and rice-belt regions.[3] Although Table VII-1 shows that freight and passenger traffic increased considerably over the period for which there are data, the average distance hauled for a ton of freight remained fairly constant from the

[2] Chosun Government General, *Statistical Yearbook,* 1935, p. 276.

[3] For a map of the Railroad System in 1925, both completed and projected, see Chosun Government General Railroad Office, *Chosen Tetsu-do Rosen* [Collected Materials on the Chosun Railroads], Seoul, 1929(?), (in Japanese), p. 196.

'teens to the mid-1930's at around 210 kilometers, while similar data show that the average distance traveled per passenger varied even less, around a rough average of 65 kilometers. Of the South Korean Provinces, only Kangwon was seriously by-passed by rail lines, and so it is reasonable to conclude that most major regions were serviced by rail and hence brought in contact with cities and ports. It is clear that the railroad was important for shipping rice, and other data show that soybeans, barley and many other agricultural products were moved by rail in considerable quantity. But why should the railroad system and the overall pattern of shipments favor Kyonggi Province and contribute to its relatively rapid growth? There is no proof that it did, but it is reasonable that farmers living near stops on the major lines close to Seoul could benefit from markets there and from other urban stimuli. For farmers at even a short distance from those stops, however, it is doubtful that the railroads made much of an impact without other auxiliary forms of transportation. The rails were therefore undoubtedly important for moving heavy and easily stored products, such as rice, beans and fertilizers, which could be slowly collected at storage facilities by slow-moving non-rail transport. But it is less likely that the railroads by themselves brought Korea's rural communities in touch with modernity. Great emphasis must be placed on the importance of secondary transport networks of roads and highways, and it is likely that they, rather than railroads, were responsible for Seoul's influence on Kyonggi's rural areas during the Japanese period. The rapid expansion of highways after 1945 certainly shows this to be the case, not only for Seoul but also for other urban-rural links in the later period.

Analysis of road statistics for the Japanese period shows that Kyonggi Province did in fact have better highway service by the 1930's. This is reflected in the fact that although Kyonggi had fewer roads of all kinds for its size than the more level provinces in the rice belt and densely populated South, it had a significantly higher level of "first class" highways than any other province. This can be seen in Table VII-2, "South Korean Roads, by Province, 1927 and 1935." Third class roads in South Korea are often little more than dried stream beds or rutted, narrow, bumpy and rock-filled ways. Second class roads are better, but only first class roads (still unpaved, of course) are of the quality needed to make bus travel and truck transport practical over any distance. The relationship between first class roads and rural growth is further confirmed when it is noticed that

Table VII-2.   *South Korean Roads, by Province, 1927 and 1935*

| | Length of Roads (km) | | | | | | Length per Province Area (km/km²) | | |
| | 1927 | | | 1935 | | | 1935 | | |
	First Class	Second Class	All	First Class	Second Class	All	First Class	Second Class	All
Kyonggi	358	260	1,436	365	266	1,588	28.4	20.7	123.7
Kangwon	108	1,140	1,836	108	1,196	2,121	4.3	45.5	80.6
N. Chungchong	111	193	782	135	234	909	18.2	31.5	122.4
S. Chungchong	183	266	914	183	284	1,096	22.5	35.0	134.9
N. Jolla	141	244	1,193	146	291	1,474	17.1	34.1	172.6
S. Jolla	138	595	1,512	132	592	1,877	9.5	42.6	135.0
N. Kyongsang	243	629	1,416	241	728	1,849	12.7	38.3	97.3
S. Kyongsang	160	550	1,359	164	579	1,790	13.3	47.0	145.2
South Korea	1,442	3,877	10,488	1,474	4,170	12,704	13.6	38.4	117.1

Sources: Calculated from road and area statistics in Chosun Government General, *Statistical Yearbook*, 1927 and 1935.

Table VII-3. Paved Roads, 1936-1975

	National and Provincial Paved Highway (km)	Share of Total Roads Paved (%)	Limited Access Expressway (km)	National Paved Roads (km)	Share of National Roads Paved (%)
1936	339	2.1	—	293	4.8
1944	827	4.9	—	746	12.2
1947	809	5.3	—	737	14.3
1950	403	2.6	—	364	7.0
1953	329	2.1	—	294	5.1
1956	369	2.3	—	332	5.8
1959	612	3.8	—	554	9.7
1961	779	4.8	—	721	12.6
1963	919	5.4	—	865	14.9
1965	1,113	6.1	—	1,042	17.7
1967	1,494	7.9	—	1,442[a]	17.6[a]
1969	1,772	9.4	472	1,652	20.4
1971	2,542	13.4	655	2,302	28.3
1973	3,358	17.5	1,013	2,869	34.6
1975	4,325	22.6	1,142	3,620	44.0

Sources: Unpublished data sheet from the Ministry of Construction (M.O.C.), compiled from the M.O.C., *Statistical Yearbook*, 1960 (for 1936-1959 data), individual years of the same yearbook (1961-66), and individual provincial road reports (1967-75).

Note: [a] There was a reclassification of roads in 1966, with over 2,000 km of provincial roads re-classified as national roads. As a result, the data are not consistent between 1965 and 1967.

the poorest province, Kangwon, has only 4.1 kilometers of first class road per square kilometer of area (in contrast to 28.4 for Kyonggi and 13.6 for South Korea as a whole). But it would be a mistake to make too much of these differences in first class roads. It is interesting that when first and second class roads are taken together, almost every province has roughly the same road service, closer to 50 kilometers per square kilometer of area. Nevertheless, when recognition of Kyonggi's higher quality roads is combined with our knowledge of Seoul's dominant position in South Korean industry, it is easy to understand the Northwest's rather special position and more plausible to assign responsibility for rural Kyonggi's divergence from the national pattern to the influence of the nation's largest urban area.

Although highway construction was quite rapid during the decade of the land survey and the early 1920's, Table VII-2 also shows that in the South Korean provinces the pace of construction had slowed considerably and was concentrated almost totally in work on third class roads. Radical improvement in the quality of existing roads in the form of hard-surface paving, however, began in the period of the 1930's, and although this factor remained of limited significance through the war years and the 1950's, it came to be of very great importance during South Korea's years of rapid economic growth of the 1960's and 1970's. The passages which follow will concentrate on the construction of paved highways during the years after World War II and in particular during the years following the 1961 army coup d'etat.

Table VII-3, "Paved Roads, 1936-1975," shows the meager level of road paving in 1936. It is also interesting to notice, however, that by the end of the War in the Pacific, the Japanese had paved more kilometers of roads than would again be paved for over fifteen years. The destruction caused by the Korean War, which ranged over all of South Korea in the latter half of 1950, is clear, for between 1947 and 1953 the length of paved roads was approximately halved. The most interesting years covered by the table, however, are those after 1963, when paving activities greatly accelerated. The share of national[4] and provincial highways paved increased from 5.4 percent to 22.6 percent,

---

[4] The Japanese re-classified all roads in 1939 into two subgroups: National and local roads. National roads seem roughly equivalent to the combined first and second class roads of the earlier classification, although some "second class" roads seem to have been re-classified as regional roads. For this reason, there is some difficulty in comparing statistics from Table VII-3 to those of Table VII-2. For an example of this break in

and when national roads alone are considered, the increase was from 15 percent to 44 percent. Of equal importance was the construction of limited access expressways, first between Seoul and Incheon, then between Seoul and Taejeon, and finally between all of South Korea's important cities.

What was the impact of this paving on South Korea's domestic freight traffic? Paving greatly increases the speed and ease with which trucks and buses can cover their routes, and the impact of paving can be seen in Table VII-4, "Domestic Freight Transportation, 1963-1975." These are the only years for which such statistics are available, and yet in the space of twelve short years the share of total domestic freight carried over highways increased from 45 percent to over 60 percent, while that carried by rail fell from 50 percent of the total to 31 percent. The residual represents a slight increase during this same period of the importance of coastal shipping. It is also interesting that although the average distance a ton was hauled by rail remained rather constant, that for highways roughly doubled between the years of 1969 and 1971, undoubtedly reflecting the opening of major sections of limited-access highway.

The scope and rapidity of the highway paving program can be seen in the four maps below, which give the paved national highways and expressways for 1963, 1967, 1971 and 1975. The most dramatic expansion in the paved system has been in the 1970's, and the recent completion of the expressway through South Jolla Province will perhaps alter the pattern of that area's retrogression in comparison to more northern provinces. In looking at the map for 1971, it is important to note that the portion of the expressway between Seoul and Taejeon runs through North Chungchong Province. This section was completed in 1969 and meant that the farmers in that area were suddenly within a few hours by truck of the previously inaccessible Seoul markets. This section of highway, and to a lesser extent the paving of national highways in that area by 1967, are in large part responsible for the rapid expansion of vegetable production and of income from vegetables that began in the late 1960's (see Chapter VI).

A more statistical presentation of the degree to which different regions have benefited from the road construction program is given in Table VIII-5, "National Paved Road Density by Province,

the statistical continuum see Chosun Government General, *Statistical Yearbook, 1941,* p. 158.

Table VII-4. Domestic Freight Transportation, 1963-1975

(Freight in millions of tons)

	Total Freight	Railway Freight	Share of Railway Freight (%)	Road Freight	Share of Road Freight (%)	Average Haulage Distance of 1 Ton	
						Railway (km)	Road (km)
1963	39.8	19.8	49.7	18.1	45.3	220	24
1965	49.0	22.4	45.6	24.0	48.9	225	21
1967	60.2	27.4	45.6	28.6	47.5	225	23
1969	95.3	30.6	32.1	56.6	59.4	239	23
1971	117.2	32.0	27.3	73.9	63.1	245	45
1973	119.5	37.8	31.6	72.0	60.2	228	44
1975	139.1	42.8	30.7	84.5	60.8	217	46

Sources: Calculated from Economic Planning Board, *Korea Statistical Yearbook*, various years, "Transportation" section.

Map VII-1.   Paved National Highways in South Korea, 1963

Map VII-2.   Paved National Highways in South Korea, 1967

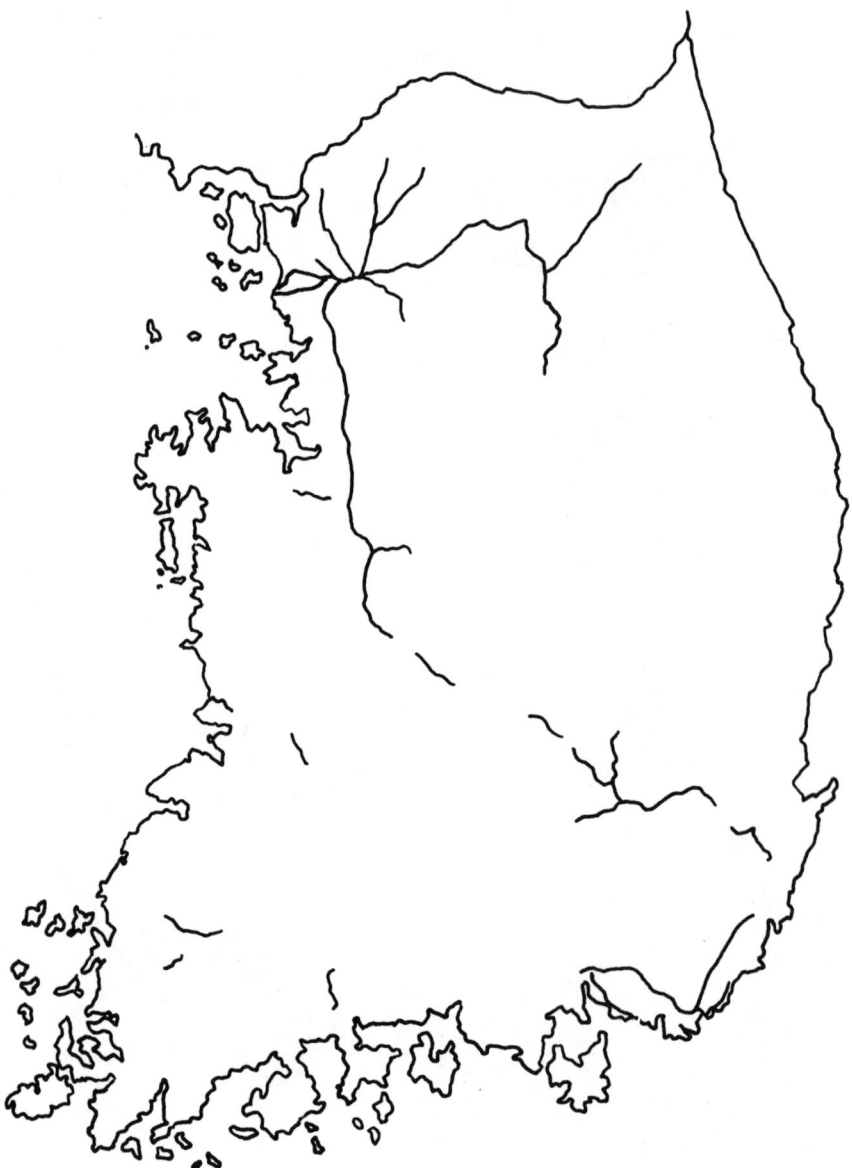

Map VII-3.   Expressways and Paved National Highways in South Korea, 1971

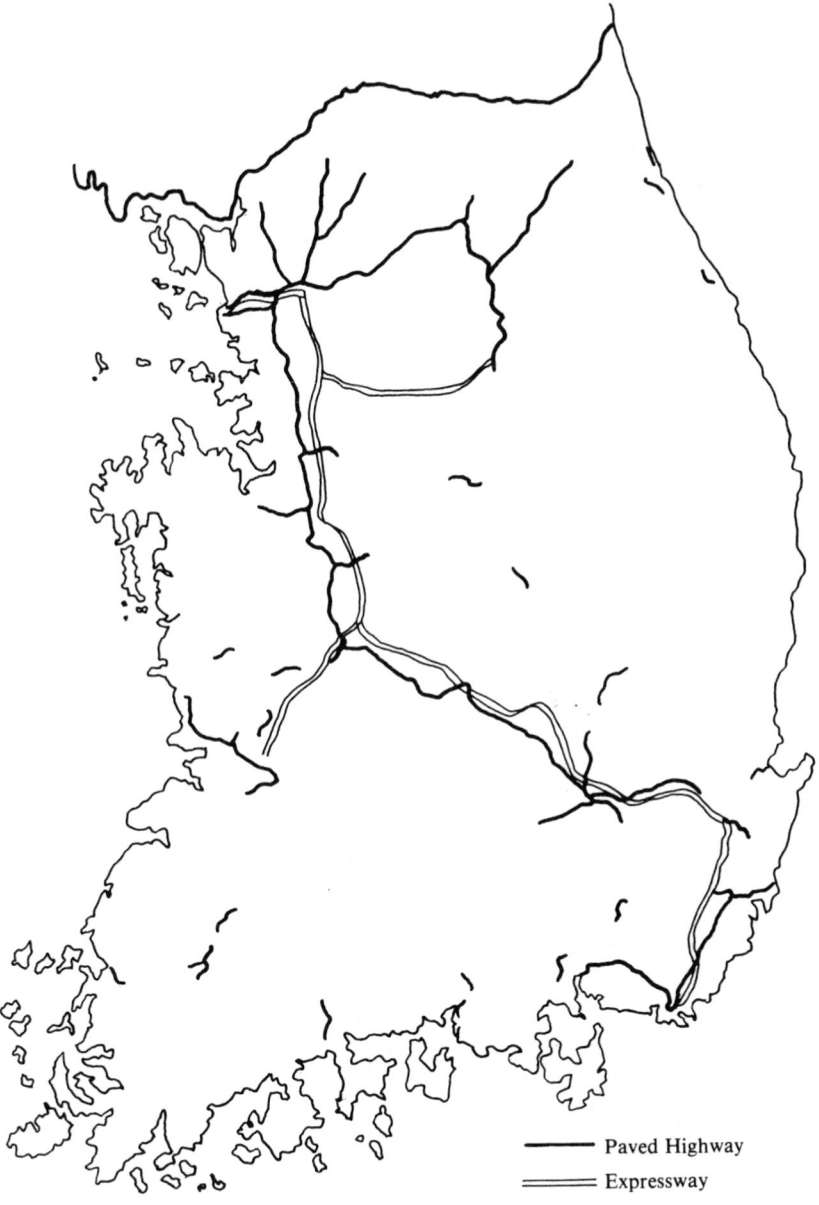

Paved Highway
Expressway

Map VII-4.   Expressways and Paved National Highways in South Korea, 1975

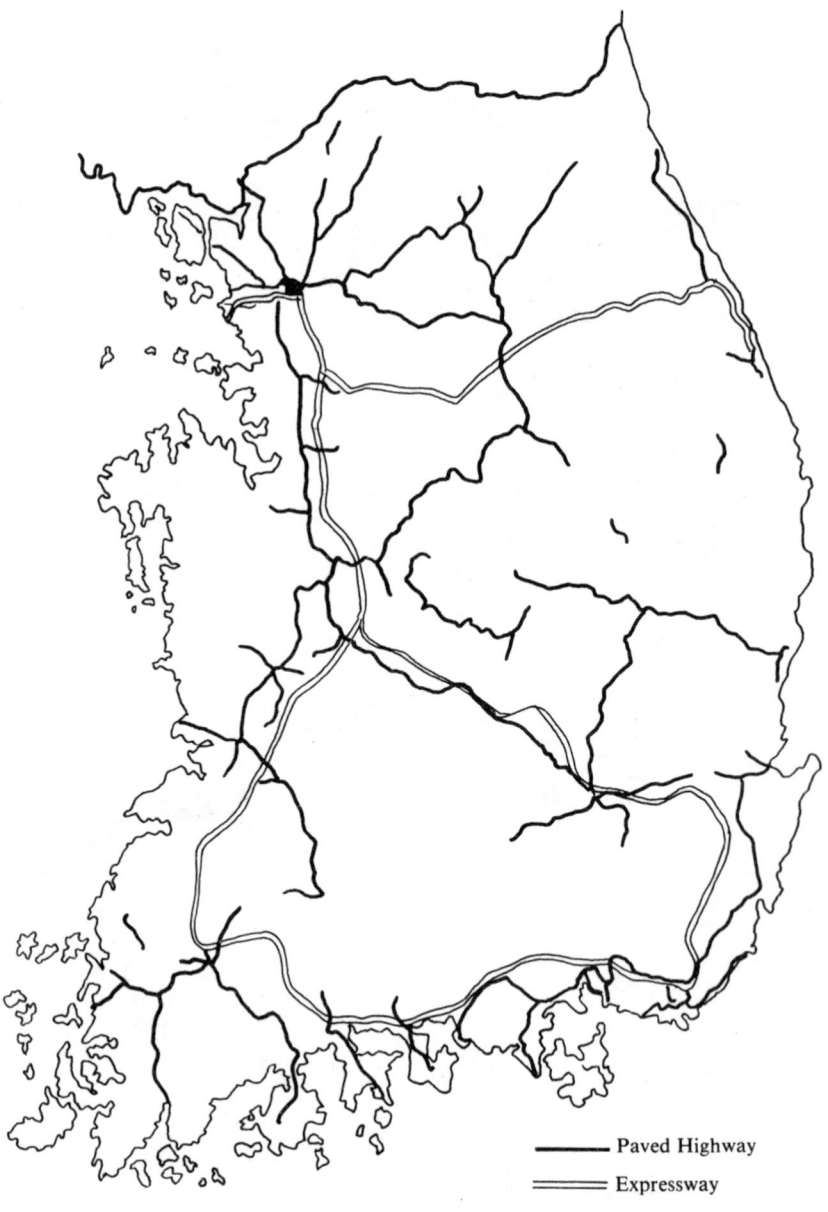

Paved Highway
Expressway

Table VII-5. National Paved Road Density by Province, 1964-1975

(km of paved road/thousand km² of area)

	1964	1967	1969	1971	1973	1975
Kyonggi	18	26	30	32	38	44
Kangwon	9	13	15	16	17	23
North Chungchong	3	10	12	19	27	41
South Chungchong	9	13	16	25	40	45
North Jolla	2	4	8	12	23	28
South Jolla	3	5	6	16	19	31
North Kyongsang	9	11	15	21	24	34
South Kyongsang	7	10	11	17	25	29
Jeju	20	34	63	130	142	142
South Korea[a]	8	12	15	22	27	35

Sources: Calculated from paved national road data in individual years of the Ministry of Construction, *Statistical Yearbook,* and total province area from Economic Planning Board, *Korea Statistical Yearbook.*

Note: [a] Does not include the special cities of Seoul and Pusan.

1964-1975,'' which shows the length of paved national highway per total area for each province. At the beginning of the period the two Jolla Provinces and North Chungchong Province had by far the fewest paved roads in relation to their total area. The statistics show, however, that while North Chungchong Province rapidly increased its density of paved highways, the Jolla's have done so much more slowly. Only since 1973 have these southern rice belt provinces been able to match the scale of paving in the more mountainous South Kyongsang region. It is also significant that the poorest region, Kangwon Province, has shown both a low level of paving and a low rate of growth in its paved highway density.

The above maps and tables make it clear, then, not only that highway paving has been dramatic in the years of the Park government, but also that the geographical pattern of paving works has benefited just those upland areas which have shown the most dramatic improvement in their relative productivity during the same period. In other words, these highway works have had a direct effect on the isolation of rural areas from Seoul and other urban-industrial centers, and that isolation in turn seems to have some indirect or direct effect of its own on rural productivity. The question remains, however, what is the nature of this rural isolation, and to what degree can and does isolation from South Korea's urban centers affect farm income?

Casual empiricism and personal experience indicate that isolation can and does make an enormous difference. The provincial data, of course, have already shown that Seoul, Pusan and Kyonggi agriculture provide a much higher level of per-capita output. Travel through rural South Korea provides a feeling for why this might be so in a country smaller than the state of Virginia.

Transportation is by bus or by rail, but for the ordinary person without reserved seats in the better buses and express trains, travel is a crowding and exhausting process. From the waiting for tickets to the rush for a place (rarely a seat), to the crush and fatigue of standing tightly packed, and finally to the long hours of slow travel over dirt roads and up inclines which settle the standing throng towards the back of the bus where less fortunate standees are pressed, bus travel on common carriers over poor roads is hard work and is undertaken only for good reasons. Where the roads are level and have been paved or on the limited access highways, the ride is faster and more comfortable, but the inconvenience of ticket purchase, of crowding and of transfer delays is still significant. Train travel is faster between stops, but is equally crowded for all but the better coaches, and the inconveniences of local stops and of course inflexible track and local station stop patterns makes it less useful for all but long trips over heavily traveled routes.

The typical Korean farmer living at some distance from a city, not to mention the farmer also living at some distance from a town with bus service, has very little chance to consider selling vegetables and other products in urban markets. The marketing process for perishable products is extremely difficult, involving as it does long distances and frequent changes in carriers. For many Korean villages access is by foot or oxcart only, and even where feeder roads have been built, the town to which it leads is many kilometers away and itself isolated from still more distant city markets.

For farms near large cities or farms near entries to the limited access highway, however, the differences are dramatic. Farmers near cities can cut their cabbages, lettuce, Welsh onions and other daily-use vegetables every morning and load them in large push-carts or rented trucks (the distances are short and the fare relatively low) and sell the day's harvest by noon. Farms near the super highway can load a truckful of cucumbers or eggplant and be in Seoul or Pusan or Taegu within hours. In addition, farmers with such direct access to cities stay closer to the trends in urban wants.

The above observations are drawn from periods of personal farm work and rural travel during a year and a half stay in South Korea. They do not represent data about isolation and farm life, but rather provide an intuitive context for understanding statistical results relating farm income to urban-industrial access. Does such a relationship exist? The measure of county isolation developed below allows us to show without question that it does. Such a measure must deal with two conceptual difficulties. First, isolation of a county from a center of industry involves both the distance to the center and the importance of the center. Some way is needed to combine information from both these sources in a single index. Second, South Korea has many more than one center of urban industrial activity. The several influences of different cities on a single county must somehow be summed in accordance with an acceptable rationale.

The problem of distance and city size is handled with concepts from location theory used to measure "access" or "interaction" between two centers of population concentration. These concepts are rooted in theories of Newtonian physics relating gravitational attraction between two bodies to their respective masses and the separating distance.[5] In general, the theories allow for greater "access" (A) the greater the sizes (P and p) of the population centers and the smaller the distance (D) separating them. This relationship can be summarized in the following expression:

$$A = \frac{a\,P^b\,p^c}{D^d} \dots\dots\dots\dots\dots\dots\dots\dots\dots\dots\dots\dots\dots\dots\dots\dots (1)$$

where a, b, c and d are a scale constant and exponents for P, p and D, respectively. For the present analysis of South Korea, "access," A, measures the number of buses traveling between the two locations in the course of a week.

The second conceptual problem of combining the influence on one county of several industrial centers (numbered j, with $j = 1, ..., n$) is solved by defining the total access of a county as the total number of buses which arrive from urban areas in the course of a week. Refer-

[5] See Walter Isard, *General Theory; Social, Political, Economic and Regional,* Cambridge: M.I.T. Press, 1969, pp. 60-61.

ring to the rural community as county i, the access of that county to urban area j can be symbolized as A_{ji}. The total access of county i from all industrial centers can be written as A_i and is merely the sum of all A_{ji}:

$$A_i = \sum_j A_{ji} \dots\dots\dots\dots\dots\dots\dots\dots\dots\dots\dots\dots\dots\dots (2)$$

Substituting the full expression for A_{ji} in accordance with expression (1), we obtain the following formula for total access:

$$A_i = \sum_j \frac{a\,P_i^b\,P_j^c}{D_{ji}^d} \dots\dots\dots\dots\dots\dots\dots\dots\dots\dots\dots\dots\dots (3)$$

where the symbols and subscripts have the following meanings:

A_{ji} is the level of access between city j and county i. In practical terms this means the number of passenger buses traveling from city j to county i in a week.

A_i is the total level of access for county i to centers of urban industrialization; that is, the weekly number of city bus arrivals.

P_i is the population of destination county i.

P_j is the population of city of origin j.

D_{ji} is the distance between city of origin j and destination i.

The first step in calculating a measure of county isolation is the determination of the parameters a, b, c and d. There were fourteen cities in 1970 with manufacturing employment over 5,000 persons,[6] and the above parameters were estimated using bus frequency and route distance data between these cities and outlying communities throughout the country for which suburban or interurban bus service existed. Population data for the cities of origin and for each of the destination communities were also collected.[7]

---

[6] The fourteen cities are Seoul, Pusan, Incheon, Suweon, Chuncheon, Cheongju, Taejeon, Jeonju, Gunsan, Kwangju, Mogpo, Taegu, Masan and Ulsan. See Table IV-5, p. 70 (especially the accompanying note), and Map IV-5, p. 64.

[7] The bus data are from Government of South Korea, Ministry of Transportation, "Intra- and Inter-Province Bus Routing, 1970," Seoul, 1971 (mimeographed, in Korean). The populations of destination communities were gathered from individual provincial yearbooks for 1970, and data on manufacturing employment were taken from Sa-hon Kim, "Data from the 1970 Mining and Manufacturing Census," undated (unpublished, in Korean).

Our theory of rural "access" tells us that the number of bus trips a week between one of the cities of origin and any community of destination should be explained to a significant degree by equation (1) above. The parameters of equation (1) have therefore been estimated by ordinary least squares, regressing the logarithm of the number of buses on the logarithms of P, p and D, since taking the logarithms of both sides of equation (1) yields the following form:

$$log\ A = a + b\ log\ P + c\ log\ p + d\ log\ D.$$

The results of the regression analysis are given below. The numbers in parentheses below the parameters are t-ratios.

$$log\ A = 7.074 + 0.223\ log\ P + 0.177\ log\ p - 1.500\ log\ D \ldots \ldots (4)$$
$$(12.6)\quad(4.5)\qquad\quad(4.8)\qquad\quad(-11.4)$$
$$R^2 = .463 \qquad \text{Number of observations} = 174$$

There are good reasons why the explanatory power of the equation expressed in the coefficient of determination is not greater, since we have not taken into account such basic factors as road quality (which varies immensely in South Korea) and the importance of communities of destination as tourist attractions. Nevertheless, the results can be considered good for cross-sectional data of this nature, and the significance of the individual parameters a, b, c and d is clear from the values of their respective t-ratios. The value and significance of the parameter d is particularly pleasing. For the Newtonian "gravity" model, its value is of course -2.0, and this value is sometimes used in location theory research. The results above are therefore consistent with a priori and commonly-held notions and provide parameters for a gravity model relationship which is of verified significance for rural Korea.

Using these estimated parameters to calculate an index of total "access" for each of South Korea's counties, however, presents us with a third conceptual problem. The formula for "access" given in equation (1) requires a value for the population of the destination community. For many rural counties in South Korea, however, the farm population is spread along a traversing national highway or distributed evenly around several local rail stops, with no single identifiable population center of relevance. It would therefore be preferable to generate an index of access or isolation independent of the rural county's size or population make-up.

Such a measure is possible if we suppose that for the counties of South Korea there is one and only one hypothetical center of urban industrialization. If there were only one city in the country, a county's isolation could be measured by its distance to that city. A method for determining the effective value of that single distance is given below.

The equation for a single county's access to such a single hypothetical urban center is expressed by the following formula:

$$A_i = \frac{a\ P_i^b P_h^c}{D_i^d} \quad \dots\dots\dots\dots\dots\dots\dots\dots\dots\dots\dots \quad (5)$$

where P_h is the size of the hypothetical single center of urban industrialization.

   D_i is the single distance from county i to the hypothetical urban center.

It should be noted that the expression above is a second formula for calculating A_i, the degree of total access for county i, the first being given in equation (3). It should be noted further that if the two expressions are equated, the variable representing the population of the destination county, P_i, cancels out, and the remaining equation can be solved for D_i. The resulting expression is given here:

$$D_i = [\frac{P_h^c}{\sum_j (P_j^c / D_{ji}^d)}]^{1/d} \quad \dots\dots\dots\dots\dots\dots\dots\dots \quad (6)$$

Formula (6) provides, in essence, a method for aggregating the many actual individual distances from one county to South Korea's separate centers of industrial activity. The aggregation is carried out within the gravity model framework introduced above, and the resulting measure of isolation, D_i, has the intuitively pleasing interpretation of the hypothetical distance to a likewise hypothetical urban center of size P_h.

Having estimated the values of parameters c and d, the estimation of actual isolation indexes for each of South Korea's counties and towns is now straightforward. The distances D_{ji} were measured from each of 167 counties and towns to each of the 14 largest centers of industrial employment, using a large 1970 map of South Korea. The

Map VII-5. Isolated Korean Counties, 1970

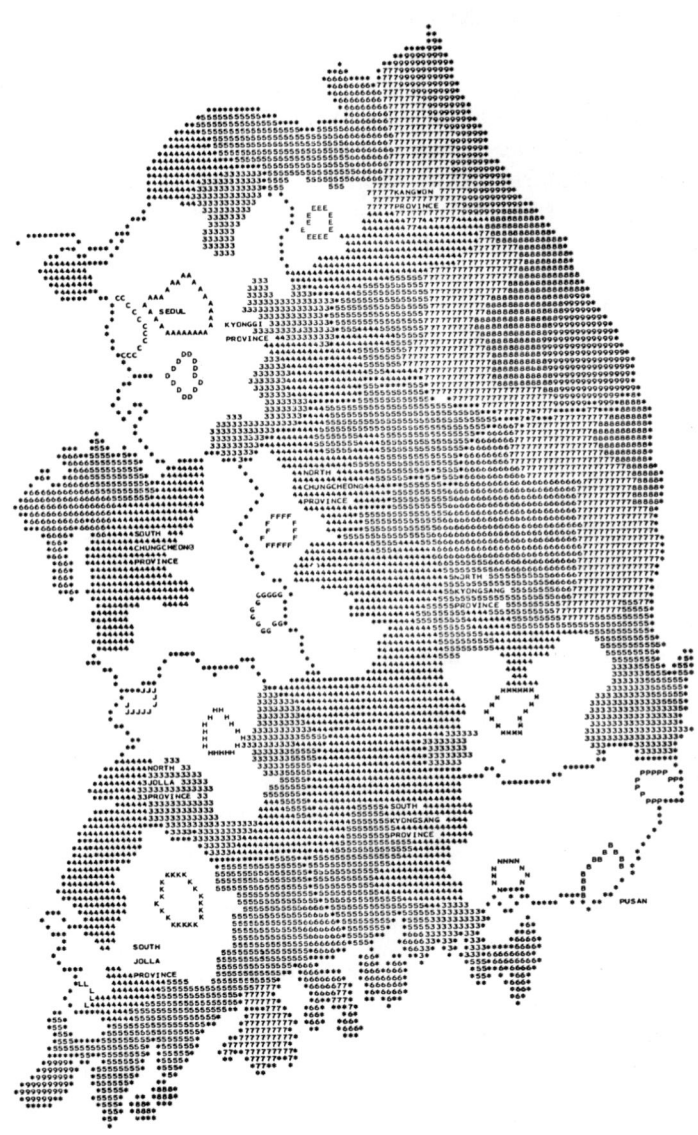

urban populations, P_j, are the same as those used for estimating the parameters of the gravity model. The population of the hypothetical center was assumed to be the same as that for a city with industrial employment of 300,000 persons, making it comparable in size to Seoul in 1970.

The resulting indexes reveal considerable variation in the isolation of South Korean rural areas in 1970. Counties near a large city or situated between two or more urban areas are predictably less isolated than counties on the perimeter of Korea's modern economic activity. The pattern of rural isolation is most easily seen by referring to Map VII-5, "Isolated Korean Counties, 1970." On this map, the more isolated counties have been "colored" in, with the higher intergers (7's, 8's and 9's) representing the most isolated regions and the lower ones the lesser isolated areas. Blank portions of the map represent counties with the most direct access to South Korea's industrial urban areas.

Reference to a second map, Map VII-6, "Poor Farm Counties in South Korea, 1970," provides an interesting comparison with the map of isolated counties.[8] The correlation is far from perfect, but in general, counties with lower per capita product also tend to be the more isolated counties. Analysis in a subsequent chapter will show that this significant relationship survives even if the counties are rendered comparable in terms of rainfall, amounts of land of different category per farm, fertilizer use and other factors influencing farm productivity.

However interesting the above results may be, it should be remembered that our measurements are only a crude estimation of the actual degree of isolation experienced by South Korean farmers. A major drawback is that quality of roads is not included in the calculation. This could have been done by using available data for speed of travel over different surfaces and basing our calculation of access on travel time rather than distance, but data on conditions of roads for 1970 were not readily available, and distances over national highway or rail were used and assumed to be of identical quality throughout the country.

A second serious drawback is that there has been no consideration of the ease or difficulty of travel within the individual counties themselves. There are many counties traversed by a national highway

---

[8] For a discussion of the measure of county agricultural product used to construct this map, see the discussion under "Farm Value-added" on page 222 below.

Map VII-6. Poor Farm Counties in South Korea, 1970

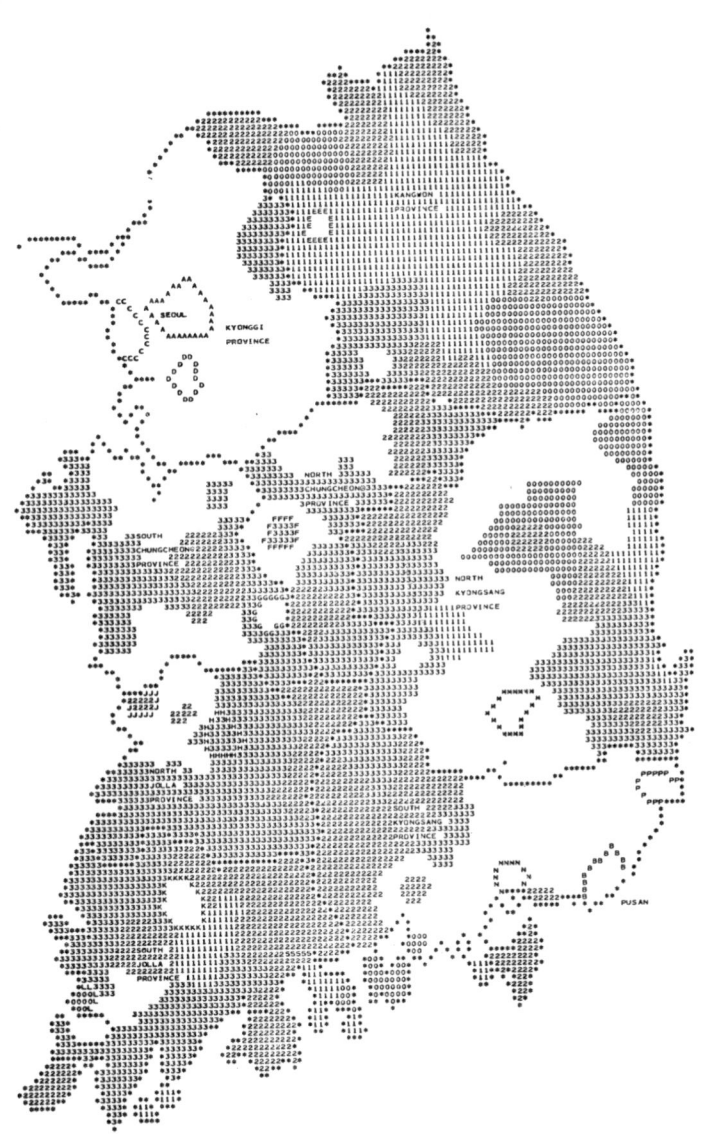

which have poor feeder roads, often because of mountainous terrain, and such isolating conditions are not reflected in our isolation indexes at all.

Finally, as mentioned above, our index is not influenced by the presence or absence of population centers in or near the counties, except for the fourteen centers listed earlier in the chapter. This shortcoming is felt in two ways. First, centers of manufacturing employment, no matter how small, reduce isolation for nearby farmers, and theoretically all locations of manufacturing employment should have been taken into consideration for our calculations. The costs of doing so, however, would have been prohibitive for this study. Second, centers of population in a county act as foci for modes of communication and access to other larger cities. A farmer close to a moderately-sized town will be effectively closer to a more distant city than a farmer at the same distance from that city but not near any town at all. To deal with this problem we would have had to determine satisfactorily the relevant population for the community of destination representing a county and then use the formula for total access given in equations (2) and (3). With no clear criteria for determining such destination population sizes, this methodology could not be used. The indexes of isolation obtained, therefore, reflect isolation only in its most global sense as the travel distances from a country's general location to Korea's major urban centers.

In spite of the shortcomings described above, the indexes probably capture a significant portion of the isolation from modernization experienced by farmers in South Korea. Counties are comparatively small geographical units, and the fourteen cities used as urban centers represent 75 percent of South Korea's total manufacturing employment. The results, therefore, provide an excellent statistical foundation for testing the major Schultz hypothesis that farm income and farm poverty are significantly related to the geographical matrix of inequality in the industrial growth of the economy as a whole.

This study's final chapter will now include this measure of rural inequality in an econometric analysis of production function relationships for rural Korea, and the results remove any doubt about the significance of this factor for rural development.

CHAPTER VIII

FARM OUTPUT PRODUCTION FUNCTIONS AND ISOLATION

This chapter provides statistical confirmation that every kilometer of distance from industrial concentration clearly reduces farm value added, even when adjustments are made for other influences, such as differences in the use of fertilizer, irrigation, and vinyl housing. To obtain this confirmation, the study considers acess to cities as just one of many ingredients responsible for farm output levels. Such an analysis of farm production as a function of numerous traditional ingredients is a common research technique. This chapter, however, uses the study's 1970 isolation index as a supplementary "ingredient" affecting 1970 county farm output levels. Readers may find the clear productivity measures for some traditional inputs interesting in their own right, but the analysis' main purpose is to provide a rigorous backdrop for isolation's unmistakable statistical influence on farm output.

Farm activity is in essence production activity. Output results from the combination of various inputs (land, labor, fertilizer and others) following a certain method which reflects the technology being used by the farmer. As input levels, input combinations and technology vary, so varies the output.

With a large enough number of separate input and output combinations it is possible to find the contribution each input makes in farm production, given that the technology used is more or less unchanged. Both Korean provincial time-series data and 1970 county data provide a large number of sample observations for basic inputs and total farm output. This chapter uses regression analysis to examine both data sets and estimates the marginal productivities of major input categories. It compares these results with the actual prices paid by farmers for the same inputs. The results confirm that uses of individual inputs is profitable and that the prices of inputs have settled at levels which to a certain degree reflect their productivity for the farmer.

The analysis also shows that when access to urban-industrial areas is included as one of the inputs, greater access makes a very significant contribution to output. Strictly speaking, access to modernizing sectors is not an input. But it represents possible differences in technology, especially differences in crop planting patterns, since choice of crops is an important part of the farmer's method for extracting value from his soil. In addition, an index for isolation may partially represent other more illusive factors contributing to output: the supply of capital, intensity of land use, higher quality of standard inputs and possibly conditions of greater human motivation. This confirmation of the overall importance of access to industrial areas is the principal finding of this chapter and perhaps of the entire study.

There are also, however, difficulties with aspects of the regression analysis presented in this chapter. In the first place, there is no clearcut reason for choosing one mathematical version of the production relationship over another. Secondly, data were not available for input variables which nevertheless have most likely been extremely important, as in the case of soil quality, for example. Thirdly, where data are available, they in some cases probably incorrectly reflect the true input they represent, as in the case of farm labor, the quantity of which probably also reflects a long-run equilibrium in population density due to regional climate and soil differences. Fourthly, for much of the data certain variables vary in such close step with others that it becomes difficult to assign responsibility for the accompanying differences in total output. In spite of these shortcomings, however, the data are numerous enough and of sufficient quality to provide a fairly detailed picture of South Korean farm activity, and the results highlighted below are important and defensible conclusions.

Finally, in addition to examining the data, weighing the forms of production relationships and analyzing the individual estimates produced, the last part of the chapter uses the same regression analysis on different regions of the country rather than merely taking the country as a whole. The results emphasize that the significance of isolation for rural development shown in this study is not just a quirk of South Korea's geography, climate and historical economic growth. The influence of access to urban-industrial centers permeates all of rural South Korea and thus deserves the attention of those also interested in rural development in other growing economies.

The first topics covered, then, are the data themselves. The amounts of land in different categories, farm labor, chemical fertilizer, pesticides, climate, blight and rural isolation are used to explain the value of farm production. Taking these variables as individual data sets, each requires special consideration, and taken as a group they present additional difficulties. The variables are discussed below, first individually and then as a composite data set.

Land: The land variables used in this study serve the double purpose of representing both original land endowment and the accumulated improvements made to that endowment by man. South Korean farmland can be divided into four categories: irrigated rice paddy, unirrigated rice paddy, upland dry fields and permanent cropland. The names of the categories are largely self-explanatory, since upland dry fields grow virtually everything but rice (in fact, there is even an upland rice variety—it is planted heavily on Jeju Island) and since permanent crop land refers mostly to orchards. Although these land categories certainly hold unique properties as land, their differing contribution to production is also the result of work done to irrigate one paddy class and the work done to create the flat floodable fields in both paddy categories. In addition, permanent crop land represents considerable investment by the farmer in trees and their care during early years of low or negligible productivity. These facts are important and introduce difficulties for the interpretation of estimates for land's importance as opposed to the importance of investment and capital.

There are some important differences between the land data obtained from provincial time series and those from 1970 county sources. In the first place, the provincial data lump permanent crop land together with upland, while the county sources provide statistics for all four categories. More importantly, however, the provincial land data are from local administrative records of land use, while the county data are from an agricultural census and hence represent the knowledge of the respondent. It is likely that the administrative land data are inaccurate, and in fact the South Korean Ministry of Agriculture and Fisheries revised its estimates of area planted to barleys beginning in

1976, but since data on regional correction factors are not available. the provincial land estimates have been used as originally reported up to 1975. It is not likely that these differences, however, are significant enough to seriously influence our results.

The absence of information about soil quality or the extent of "double-cropping" is more worrisome. It is clear that land naturally more fertile or more extensively double-cropped will produce greater annual yields in value terms if other factors remain comparable. A winter crop of barley certainly adds to a given hectare's productivity, and the fame in South Korea of land in certain northerly counties (Kimpo County in Kyonggi, for example) which produces more and higher quality crops than others attests to the significance of good soil.

But what difference does it make if we have not used data on soil quality or double cropping? It is of course regrettable that we cannot derive estimates of the relative importance of these factors for production, but the real difficulty is that we may assign the contributions made by these ghost variables to our estimates of the marginal productivities of inputs for which there are data. More specifically, we have noticed in an earlier chapter that, historically, more productive land has tended to come to support a denser farm population. In other words, the availability of farm labor for a given area of land as measured by the overall farm population is to a certain degree the result of that land's inherent productivity. If we use potential farm labor as a substitute for actual work done on farms, we may be in fact measuring the importance of both actual labor and land's unmeasured productivity. This is, however, more a problem of our labor variable and will be dealt with separately below.

Finally, it should be noted that land is in a sense a proxy for all other fixed capital as well. In other words, the data used below do not explicitly include the inputs from services of draft animals, farm buildings, farm implements and machinery, trees, service roads and pathways, or electrification. Their omission is regrettable and is just one example of the restrictions imposed on the study by time and resource limitations.

Labor: As mentioned above, the analysis here used total farm population per province or county as a proxy for labor inputs. The first and most obvious objection to this procedure is that a person living for a year in a farm community is quite different from a person working full-time for a year in that community. In fact, household surveys of rural South Korea show that in 1970 the average number of persons in a farm household was equivalent to one man working full-time for 150.7 days and one woman working full-time for 72.9 days.[1] On average for the entire country, then, a household of six persons contributes far less than the labor of six full-time laborers. This is clearly because children, students and the elderly form a considerable share of the population in all farm areas. These same data on average labor expenditure, however, also provide a basis for the partial correction for this problem and for the conversion of annual average marginal productivity per farm person to equivalent daily value contributions per male worker and per female worker. This conversion is of course open to the criticism that the above average labor contribution rates are significantly different between regions, and that source of error does indeed remain.

A potentially more serious threat to the accuracy of the estimates of labor's value contribution to farm output is the above-mentioned degree to which farm population density may reflect intrinsic qualities of soil and climate. For if land is naturally more fertile where farm population is more dense, the value for labor productivity will be biased upward, since it will reflect both the productivity of the region's additional labor input and the productivity of the region's better land and weather. It is not easy to correct for a source of bias so potentially serious for South Korea, and we should remember this difficulty when interpreting the regression results below.

Finally, there is some difficulty with the measurement of the farm population variable itself. For the provincial series, farm population is based on administrative registration of persons by family occupation in each local district of the country. When there is little movement of population, these registers may yield a fairly accurate measure of farm population, but if there is population movement as great as it appears to have been in the 1960's and 1970's, the significance of persons moving location and changing occupation without changing their registration can become important.

Comparison of administrative and census data for the two census years (1960 and 1970) reveals that in each year the administrative data overestimated total farm population and that the overestimation was

---

[1] See Ministry of Agriculture and Fisheries, *Report on the Results of the Farm Household Economy Survey, 1975*, Seoul, 1976, pp. 84-85.

more serious in 1970 (5.5 percent) than in 1960 (2.2 percent). The overestimation was, however, more regionally uniform in 1970 and in general a little higher for the rice-belt provinces than elsewhere.[2] The important question, of course, is that if the census figures are a more accurate measure of actual farm population, how serious is the error measurement thus revealed in the provincial data for the results of our regression analysis of labor's contribution to output?

Regression analysis obtains the bulk of the justification for its results from the change or difference in input levels between provinces and from year to year. The less serious level of the relative overestimation as opposed to the absolute level permits confidence in the results we will obtain with the administrative data, deficient though they be, especially when the relative size of the errors in measurement are compared to the great differences in provincial farm population figures responsible for a large portion of our analytical results.[3]

In sum, the measures for farm population used for this chapter's analysis are fraught with potential sources of bias and error, but an examination of each of the difficulties reveals that their use in our analysis can produce results which are both significant and useful. Problems which relate to the other variables in our analysis are less serious and will now be briefly dealt with.

Chemical Fertilizer: The application of chemical fertilizers represents perhaps one of the most rewarding uses of modern inputs in Korean agriculture, and though there are no data on actual consumption of fertilizer, there is a great deal of information on its sale by region and county. Given the South Korean government's nearly complete monopoly on the sale of chemical fertilizer, the pattern of farm purchases is a high quality substitute for the pattern

---

[2] The actual degree of provincial overestimation in each year is as follows ($\frac{\text{admin-census}}{\text{census}} \times 100$):

	Seoul	Pusan	KK	KW	NC	SC	NJ	SJ	NK	SK	JJ	Total
1960	−5.7	—	0.6	−7.5	1.4	7.0	−0.9	4.0	3.9	1.8	7.8	2.2
1970	8.1	6.3	4.9	3.5	5.1	6.5	5.0	7.0	4.6	5.6	7.1	5.5

For 1970 no administrative figures were published and the figures used in the above calculations represent the average of 1969 and 1971 administrative data.

[3] See Footnote 2 above.

of farm consumption. One obvious potential source of error is the possibility that fertilizer bought in one region or time period is in fact consumed in another, but although this problem may be more serious for smaller counties than for provinces, it is still not likely to present serious difficulties.

There are other problems, however, which are specific to the county and the provincial data. The provincial series from 1955 to 1975 spans the early 1960's when the Korean government changed the basis for the "fertilizer year" from an August-to-July one to a January-to-December one. There was, in addition, no available information to help render data from the former period comparable to that in the second. A review of monthly consumption rates for the latter period shows that roughly 70 percent of the year's fertilizer application is in the months from January to August,[4] and hence the error introduced by using data from the old fertilizer year prior to 1962 comes down to the error involved in substituting the previous year's fall consumption for that of the current year. The bulk of a year's fertilizer application is at rice planting and transplanting time, however, and because these data are reflected accurately in the pre-1962 data, all that can be hoped is that the variation from year to year and from province to province in this major share of fertilizer consumption will render insignificant the variation from year to year in the autumn consumption level for any one province. For 1962 and after, of course, the annual figures for fertilizer consumption exactly coincide with data for annual output.

The difficulty specific to county fertilizer consumption in 1970 is of a different nature. While provincial fertilizer data are made widely available, those for counties are not, and the data had to be gathered from provincial statistical yearbooks for 1970, and where fertilizer sales were not reported by county in these sources, individual county statistical yearbooks for 1970 were used, though these vary immensely in apparent quality. Finally, there were thirteen counties for which no fertilizer sales data at all were available. In these cases, sales were crudely estimated by multiplying the county's area of cultivated land by the average per-hectare sale of fertilizer in adjacent counties. Our measure for county fertilizer consumption is therefore imperfect, but for most of the sample the sales figures allow some idea of the degree

---

[4] See the Ministry of Agriculture and Forestry, *Yearbook of Agriculture and Forestry Statistics,* and *Report on the Results of the Farm Household Economy Survey for the 1960's and 1970's.*

to which farmers in that region have learned to use this important modern input.

It should also be noted that similar data on county fertilizer consumption were apparently used explicitly in calculating the county farm value added data used as the dependent variable. This statistical procedure probably robs econometric measures of fertilizer productivity of much of their significance. See the discussion on county farm value added data below.

Pesticides: A second important category of modern chemical inputs represents fungicides, herbicides and pesticides and will be lumped here under the heading of pesticides. For South Korean provinces from 1955 to 1975 administrative data report the total level of chemical application by weight, but comparable data for counties in 1970 are of such mixed quality and so incomplete that they cannot be used. In place of such a direct measure, data on the number of sprayers, both hand and power driven, were obtained from the 1970 Agricultural Census and used to represent the intensity with which pesticides were being applied in the various counties at the time. The proxy's usefulness depends on the variation in quantity of pesticide applied by each sprayer, but it is likely that for county averages of pesticide used per sprayer this variation is not great enough to distort the information conveyed by the between-county variation in sprayer density.

Climate: Weather is obviously of very great importance for explaining variations in farm output and in helping separate out the other factors of production which are more under man's control than the weather's. But weather is also a notoriously difficult factor to quantify neatly.[5] This is particularly true when we are trying to account for weather changes affecting the aggregate output of scores of crops, each with different characteristic planting and harvest times. For this analysis we have concentrated on weather as it affects rice, the single most important crop in South Korean agriculture.

Two principal components of weather are temperature and rainfall. Data reflecting temperature, such as number of frost-free days or mean temperature for certain months, were not available for a large enough number of counties in 1970 nor for the early years of the

---

[5] For example, see Lawrence H. Shaw, "The Effects of Weather on Agricultural Output: A Look at Methodology," *Journal of Farm Economics,* Vol. 46, No. 1, February 1964, pp. 218-230.

1955-1975 provincial series and so could not be used. Rainfall data by month for almost every county in 1970 and for areas roughly equivalent to provinces for all of the provincial series were, however, available.[6] For fourteen counties in 1970 for which no individual rainfall data were available, the averages from adjacent counties were taken.

The single most significant period for rainfall in the rice growing cycle in South Korea is the month of July. This is after the period of transplanting when water supply is also crucial, but interviews with farmers and agricultural experts in South Korea indicate that ponds and irrigation have freed South Korean agriculture from strong dependence on rainfall in the earlier months of May and June when stored water is generally available. If adequate rains do not come in July, however, after the transplanting, then there is seldom enough water left for irrigation purposes, and the dryness in that critical growing period can seriously affect yields at harvest time several months later. Statistical testing of the significance of rainfall for individual months on subsequent yields confirms this fact, and hence we have used total rainfall for July in each county or province as our only weather variable.

This completes the discussion of variables common both to the provincial time series and the 1970 county cross-section data. For each of these data sets, however, there are variables not available for the other, and these will now be introduced briefly. For the provincial series these are dummy variables for years of blight while for the county data they are the area of vinyl green-housing and the degree of isolation from urban-industrial centers.

Blight: For two years in the 1955-1975 period there was such destruction by pests and disease that output in every province was seriously affected. The more serious of the two blights was the result of a "red rust" disease which struck the 1962 barley crop, but a rice blight in 1965 was also very damaging, and so dummy variables for every province for these years have been introduced in the analysis.

Vinyl Greenhouses: In the late 1960's and the 1970's the use of clear vinyl plastic spread over arching wooden frames and semi-

---

[6] The county data are from provincial and individual county statistical yearbooks. The annual series by region were copied from the hand-written records at the Central Meteorological Station in Seoul.

permanent wooden structures to form hothouses rapidly became more and more widespread. The vinyl houses allow earlier planting of seedbed rice shoots, and more importantly, allow vegetable planting and cultivation in early spring and late fall; in some areas, through the use of double vinyl houses (one inside the other), lettuce and other vegetables are planted throughout the winter. The vinyl house technology, then, provides a method for altering the effective climate of the land, and in a country with winters as severe and prolonged as those in Korea, the impact has been understandably very great. The data on area covered by vinyl greenhouses were gathered by the 1970 Agricultural Census and require no further interpretation.

Isolation: This variable is different from any examined in the provincial analysis, and its description and interpretation have been given in detail in the preceding chapter. To summarize, this variable gives the effective distance by national highway or rail from the county in question to a single hypothetical city employing 300,000 persons in manufacturing (a city roughly equivalent to Seoul). The significance of this variable can be interpreted in a number of ways, including market access, access to credit and technology, availability of non-farm employment and the presence of stimuli for learning and changing individual tastes. These interpretations will be touched on briefly in the next and concluding chapter.

This completes the discussion of the independent variables (the inputs) used in this chapter's analysis. It remains only to introduce the dependent variables (the outputs) before discussing problems with the data sets as a group.

Gross Farm Product: For the provincial time-series data, farm output is measured by taking individual crop levels by weight, multiplying by that crop's price in 1970 and summing to obtain a value figure for overall farm production. The measure obtained includes the production of livestock,[7] but not income from activities done off the farm. Hence, the measure of output is purely one of output from land and man's ministerings to the land. The results are sensitive to the relative crop prices of the base year chosen, but the final value obtained is nevertheless a good measure of total annual output in each South Korean province.

---

[7] See Appendix B for a discussion of how this has been done.

Farm Value Added: For the 1970 county cross-section analysis, value added rather than gross output was used, because its measure by county was undertaken by South Korea's Home Ministry for the years 1967-1970[8] and because the measure of output included income from sources other than farming. Although this is a distortion of actual output from the land, if we are trying to explain the total productivity of a farm household with variables which include the isolation from urban areas, a more comprehensive measure of production is suitable.

The Home Ministry calculation of value added, subtracting costs from total value of output, is not well documented, and conversations with Ministry officials have revealed that fertilizer and seed were probably the only cost elements considered. The Home Ministry oversees the compilation of county statistical yearbooks, and the value added calculations were apparantly made by hand from current value calculations for each crop, less the above-mentioned costs. As such, the methodology was quite similar to that employed in the present study to calculate annual provincial gross farm output, with the obvious difference of fertilizer and estimated seed cost deductions for the county data. The fertilizer cost deduction should be remembered when interpreting the inconsistent statistical significance of fertilizer in explaining value added variations. This completes our discussion of the dependent variables used in the study, and it also completes our discussion of individual data elements.

Just as we need to make special considerations for our data as separate variables, so is it important to recognize that as a composite group of data there can also be problems worth watching. For the data used here these problems are reflected in the multicollinearity and heteroskedasticity revealed by the regression analysis.

If available data presents variation which is similar for two or more variables, it is difficult to know which of these is responsible for any accompanying variation in output or to what degree each is important. Such problems of collinearity can be particularly serious for agricultural production functions where levels of inputs such as chemical fertilizer vary with the availability of certain types of land and with the degree to which that land is irrigated. The data used in this chapter's analysis is no exception in this respect, and as a result,

---

[8] See Ministry of Home Affairs, "Value-added by County in Farming, Fishing and Forestry Households, 1967-1970," Seoul, 1972 (mimeographed, in Korean). The study used National Accounting methodology to arrive at its estimates.

although the explanatory power of some equations seems great, that of individual variables cannot be determined. The problem is most serious for the provincial pooled time-series, but presents difficulties for the county analysis as well. Although additional data to ease the problem are not available, it is nevertheless possible to break the county sample into sub-regions, and we will see at the chapter's end that this indeed offers us some relief, since the multicollinearity is much more severe in the rice-belt regions than where the mixture of land types and inputs is less uniform.

The second problem with our data is the result of the enormous variation in the size of counties and provinces. The correct interpretation of regression analysis results depends on our inputs explaining all but a random portion of the variation in the output measure. If the unexplained portion is not random, and in particular if it is larger for larger observations than for smaller ones, there is a high probability that the results will be biased in one direction or another. Although this is only true for estimates of the precision of our results, it is nevertheless worthwhile attempting a correction, and this is also done at the end of the chapter when the county data are converted to per-household quantities, thereby removing the influence of variation in county size. This method also has an interesting and predictable effect on the estimate of labor's contribution, given the discussion above pointing to the confusion between farm population (and number of households) on the one hand and regional climate and soil productivity differences on the other.

These then are our data. They are far from perfect, but they nevertheless clearly hold promise of yielding interesting results. What kinds of questions can we hope to answer from these sources? In the first place, we need to know the degree to which these inputs and other factors succeed in accounting for the variation from year to year and region to region in the value of farm production. Have we in fact isolated the sources of Korean farm output? Secondly, we are interested in the actual value estimates of the contribution of each individual input. Do they conform with the cost of these inputs to the farmer? Thirdly, we are very curious about the significance and value contribution of the factors summarized in the index of isolation from urban industrial centers. Is isolation statistically relevant, and if so, just how important is it? Finally, if by breaking up the sample into its regional components we can still come up with significant results, what additional information about the regional nature of Korean agriculture can we discover? For example, is the significance and magnitude of isolation's impact consistent from one region to the next?

Before proceeding to the results of our analysis, it is important to say something about the specifications of the production function relationship we will use. Theoretically, it is not obvious that one algebraic way of relating these inputs to output is better than another. One important consideration is the degree to which the specification chosen allows for diminishing returns to increased amounts of any single input. It is clear that greater and greater amounts of, say, chemical fertilizer will make a smaller and smaller contribution to output if land, irrigation, seeds, and other inputs are not also increased. A production function specification which allows such individual diminishing returns is the log-log (Cobb-Douglas) specification, by which the logarithm of farm output is regressed on the logarithms of individual inputs. The marginal product of each input at average levels of inputs and outputs can be calculated using the formula for the marginal product of the ith input in a Cobb-Douglas production function:

$$\frac{dY}{dx_i} = \frac{a_i \bar{Y}}{\bar{x}_i}$$

where Y is output, $\bar{Y}$ is average output, x_i is the ith input, $\bar{x}_i$ its average value, and a_i the ith coefficient from the log-log specification.

If diminishing returns to scale are not considered particularly important, a straightforward linear specification gives estimates of marginal products for inputs as well. In this case, output is regressed directly on inputs, and the values of the coefficients obtained are themselves the marginal product estimates. Both of these specifications, the direct linear and the log-log, are used below, and the results from each contribute to our knowledge about Korean agriculture.

Regression Results: Although the most significant results show the independent importance of isolation for influencing farm income, the findings for traditional farm production inputs are also interesting in their own right. This is in spite of difficulties with both the nature of the data set itself and the forms of the causal relationship tested. For provincial data, the land variables are too closely correlated with one another to give completely separable results, but fertilizer shows its considerable contribution, and the labor productivity estimates are of

Table VIII-1.   Provincial Regression Variable Names and Summary Statistics

Symbol	Meaning	Units (1970 won)	Mean	Standard Deviation	Coefficient of Variation
GFP	Gross Farm Product	10^9 won	51	36	1.42
IRR	Irrigated Paddy	10^3 hectares	149	104	1.43
NONIRR	Non-irrigated Paddy	10^3 hectares	42	32	1.31
UPL	Upland	10^3 hectares	86	48	1.79
FPOP	Farm Population	10^3 persons	1,378	901	1.53
FERT	Chemical Fertilizer	10^3 tons	40	36	1.11
PES	Pesticide	tons	1,865	2,443	.76
RAIN	July Rainfall	millimeters	301	146	2.06

RED Dummy variable of 1's for 1962 to account for that year's Red Rust blight in barley.
D65 Dummy variable of 1's for 1965 to account for that year's rice blight.

Table VIII-2.   Provincial Regression Results with a Linear Specification

Variable	Coefficient	*t*-ratio	Units (1970 won)
GFP	(dependent variable)		
Constant	− 10.8	− .7	billion won
IRR	79.7	3.4	thousand won per hectare
NONIRR	44.7	.8	thousand won per hectare
UPL	40.1	1.9	thousand won per hectare
FPOP	11.6	5.3	thousand won per person
FERT	444.9	9.1	thousand won per ton
PES	185.2	.5	thousand won per ton
RAIN	4.1	1.1	billion won per millimeter
RED	− 79.0	− 3.6	billion won
D65	− 24.2	− 1.1	billion won

$R^2 = .964$ Number of Observations = 223[a]

Note: [a] The observations consist of 21 years (1955-75) for each of the 8 mainland provinces, 21 years for Jeju Province and Seoul Special City, and 13 years (1963-75) for Pusan Special City, bringing the total number of pooled cross-section and time-series observations to 223. Before 1963, Pusan was statistically part of South Kyongsang Province.

Table VIII-3.   Provincial Regression Results with a Log-log Specification

Variable[a]	Coefficient	t-ratio	Marginal Product[b]	t-statistic value[c]
LGFP	(dependent variable)			
Constant	4.57	13.6	—	—
LIRR	− .020	− .3	− 7.0	27.3
LNONIRR	.054	.8	64.9	84.1
LUPL	.317	6.3	187.4	29.6
LFPOP	.494	6.0	18.3	3.0
LFERT	.145	3.7	185.2	49.8
LPES	.110	5.8	3,023.4	522.2
LRAIN	.080	2.7	13.6	5.0
RED	− .294	− 4.0	− 15.0	3.8
D65	− .072	− 1.0	− 3.7	3.8

$R^2 = .974$ Number of Observations $= 223^d$

Notes: [a] Variable names are the same as those presented in Table VIII-1, except that the prefixed L indicates the natural logarithm of the variables.

For this log-log (Cobb-Douglas) specification, $y_i = a_o \pi_j x_{ij}^{a_j} + \varepsilon_i$, the marginal product at $\bar{x}_j$ (all j), is $\frac{\delta y}{\delta x_j} = \frac{a_j \bar{y}}{\bar{x}_j}$

[b] For units see the units for corresponding coefficients in Table VIII-1.

[c] For units see the corresponding values in Table VIII-1.

[d] The observations consist of 21 years (1955-75) for each of the 8 mainland provinces, 21 years for Jeju Province and Seoul Special City, and 13 years (1963-75) for Pusan Special City, bringing the total number of pooled cross-section and time-series observations to 223. Before 1963, Pusan was statistically part of South Kyongsang Province.

Table VIII-4. County Regression Variable Names and Summary Statistics

Symbol	Meaning	Units	Mean	Standard Deviation	Coefficient of Variation
FVA	Farm Value Added	million won	3644	2160	.593
IRR	Irrigated Paddy	hectares	4721	3344	.708
NONIRR	Non-irrigated Paddy	hectares	2449	1761	.719
PERM	Land in Permanent Crops	hectares	642	549	.855
UPL	Upland	hectares	4701	3137	.667
FPOP	Farm Population	persons	83920	46237	.551
FERT	Chemical Fertilizer	metric tons	3458	2112	.611
SPR	Pesticide Sprayers	units	4455	2977	.668
VINYL	Area of Vinyl Greenhouses	hectares	2219	5277	2.378
RAIN	July Rainfall	millimeters	625	159	.254
ISOL	County Isolation	kilometers	35	19	.529

Table VIII-5. *County Regression Results with a Linear Specification*

Variable	Coefficient	*t*-ratio	Units
FVA	(dependent variable)		million 1970 won
Constant	207.4	1.00	million 1970 won
IRR	281.8	7.62	thousand won per hectare
NONIRR	215.6	4.61	thousand won per hectare
PERM	201.4	2.03	thousand won per hectare
UPL	91.8	3.75	thousand won per hectare
FPOP	9,218.2	2.55	won per person
FERT	− 61.3	− .81	thousand won per metric ton
SPR	99.2	2.53	thousand won per sprayer
VINYL	38.2	3.95	thousand won per hectare
RAIN	629.9	1.48	thousand won per millimeter
ISOL	− 8.6	− 3.32	million won per kilometer
$R^2 = .937$		Number of Observations = 167	

Table VIII-6. *County Regression Results with a Log-log Specification*

Variable[a]	Coefficient (i)	*t*-ratio (ii)	Marginal Product (iii)	*t*-statistic value[b] (iv)
LFVA[c]	(dependent variable)			
Constant	− 1.694	− 3.26	—	—
LIRR	.116	2.44	89.5	36.7
LNONIRR	.093	2.46	138.4	56.3
LPERM	.053	2.46	300.8	122.3
LUPL	.099	2.13	76.7	36.0
LFPOP	.329	3.26	14,290	4,383
LFERT	.082	1.17	86.4	73.9
LSPR	.210	3.51	171.8	48.9
LVNAL	.006	.64	9.9	15.4
LRAIN	.051	1.00	297.4	297.4
LISOL	− .091	− 4.35	− 9.5	2.2
$R^2 = .951$		Number of Observation = 167		

Notes: [a] For this log-log (Cobb-Douglas) specification, $y_i = a_o \prod_j x_{ij}^{a_j} + \varepsilon_i$, the marginal product at $\bar{x}_j$, $j = 1 \ldots m$, is

$$\frac{\delta \bar{y}}{\delta x_j} = \frac{a_j \bar{y}}{\bar{x}_j}$$ For units, see the corresponding coefficients in Table VIII-1.

[b] Obtained by dividing column (ii) into column (iii).

[c] The variable names are the same as those presented in Table VIII-1, except that the prefixed *L* indicates the natural logarithm of the variables in question.

an order of magnitude which compares favorably with rural wages. The 1970 county data give somewhat different results for different equation forms, but for the country as a whole they also give reasonable estimates for the productivity of labor, while at the same time showing that pesticide spraying and land contribute in degrees related to their prices. The most important results, however, measure the contribution of access to modern industrial centers. Even when the county data are broken up into regional sub-groups, and even when all variables are used in their per-household form, isolation emerges as the most consistently significant factor influencing output.

Tables VIII-1 through VIII-6 provide definitions and results for regressions using both the linear and the log-log specifications. The first three tables treat the provincial time series data from 1955 to 1975, and the second three tables present results for the 1970 county cross-section sample. Note that the 223 observations for pooled provincial data include 21 observations for Seoul Special City and 13 for Pusan, in addition to 21 observations for each of the 9 provinces. Four equations have been estimated in all, then, and in all four cases the overall explanatory power as reflected in the coefficient of determination (R^2) is quite high. For the pooled provincial data the significance of the linear specification (.964) is slightly lower than that for the log-log specification (.974), and the same pattern holds for the linear (.937) and log-log (.951) treatment of the county data. In all four cases it is clear that the data marshalled explain a considerable portion of the overall variation in farm output.

When the estimates for individual coefficients are examined, however, the results are more mixed. For the provincial data, the results for the categories of land are particularly disappointing, although much more so for the log-log specification than for the linear one. A major reason for this is revealed in Table VIII-7, "Selected Provincial Correlation Coefficients," where the extremely serious nature of the multicollinearity between land variables and farm population is evident. In particular, the correlation (.986) between irrigated and non-irrigated paddy makes it very difficult for the log-log specification to separate the contribution of the two land categories. A comparison of the significance (t-ratios) of other individual coefficients for the linear and log-log specification reveals that although chemical fertilizer and farm population are very significant in both equations, pesticides appear significant only in the log-log model. Again, the reasons for this instability are suggested by the

Table VIII-7.   Selected Provincial Correlation Coefficients

Variable	IRR	NONIRR	UPL	FPOP
IRR	1.000	.883	.835	.948
NONIRR	.883	1.000	.726	.881
UPL	.835	.726	1.000	.866
FPOP	.948	.881	.866	1.000

	LIRR	LNONIRR	LUPL	LFPOP
LIRR	1.000	.986	.747	.899
LNONIRR	.986	1.000	.701	.871
UPL	.747	.701	1.000	.739
FPOP	.899	.871	.939	1.000

Note: For the meanings of the column and row names see Table VIII-1 on page 160.

collinearity in the sample, for pesticide application and use of chemical fertilizers are quite closely correlated (.801). In discussing the pesticide results, it is also useful to note the historical pattern of pesticide use. Examination of monthly data and discussions with specialists reveal that heavy pesticide and herbicide applications resulted from bad harvests in the early 1960's. Because of blight, herbicide use was greater. Hence, for those years, and for the regions affected most severely by disease, there is a negative correlation between chemical applications and crop output levels. In other words, the results for individual coefficients are mixed because the nature of the pooled provincial data make it difficult to obtain consistent results, though the values of individual coefficients are still interesting and worthy of comment below.

A most general criticism of both the linear and the log-log regressions points to the high probability that the equations have either ignored variables that are important (e.g. soil quality and numerous weather variables) or ignored other causal relationships (e.g., between climate, soil and long-run population density patterns). That is, causal relationships have probably been misspecified in a number of dimensions, largely because of the shortage of useful data and the shortage of resources for seeking them. The individual results, therefore, are most likely biased from the "true" (but unobservable) values, resulting in the different coefficients for similar variables in different equations.

The curious reader might also have noted that the differences in results between linear and log-log methods in a sense reflect the overall influence of the logarithm operation, which tends to reduce a high datum value by a greater proportion than a low value. Hence, for irrigated paddy in the provincial data, for example, the considerable variation between the land-rich South-west region and the others would be de-emphasized in the log-log specification, as in the sample a crucial degree of variation is needed to isolate the marginal productivities of different categories of land. These criticisms are particularly serious for the provincial pooled time series and cross-section, in part because of the only slight variations in land use for a given province over time.

In contrast to the provincial results, comparisons of log-log and linear coefficients for the 1970 county data, however, show a more interesting pattern. The linear specification is in general better for obtaining estimates of the contribution of land and vinyl greenhouses, while the log-log specification is more successful in estimating coefficients for farm population, fertilizer and pesticide sprayers. When it is noticed that land of all categories is in short supply in South Korea and that vinyl houses are still relatively recent in their introduction, it is reasonable to conclude that these inputs are far from levels where diminishing returns are relevant, and hence the linear specification is more suitable. Farm labor, chemical fertilizers and pesticides, however, are all examples of inputs in ample supply in South Korea, and it is likely that their use is at levels much closer to where diminishing returns might set in, making the log-log specification much more suitable for estimating their marginal contribution to production.

In any event, in spite of the difficulties introduced by the strong collinearity in both samples (for the county data fertilizer is strongly collinear with the four land categories and may have been used explicitly in the calculation of the value-added dependent variable), many of the individual results remain extremely significant, and it is interesting to see the degree to which the estimated marginal productivity of each corresponds to actual prices for those goods. Table VIII-8, "Prices of Selected Agricultural Inputs, 1959-1975," provides a basis for making such comparisons.

As noted above, the provincial estimates of land's contribution to output are disappointing, and the low value in comparison to the much higher quality estimates from county data further attests to the

Table VIII-8. Prices of Selected Agricultural Inputs, 1959-1975

(1970 won)

| | Farm Land (1000 won/ha.) | | Labor | | | Chemicals | |
	paddy	upland	male (won/day)	female (won/day)	farm person[a] (won/year)	sprayer (won/unit)	fertilizer (won/kg.)
1959	543	288	391	242	12,933	6,633	112
1960	637	349	361	222	11,923	6,639	114
1961	717	374	369	226	12,176	6,805	108
1962	809	405	362	220	11,924	7,585	107
1963	911	437	405	258	13,487	7,637	99
1964	899	414	444	277	14,714	6,357	96
1965	930	441	427	272	14,219	6,558	118
1966	902	437	435	284	14,571	6,497	106
1967	851	439	467	315	15,767	6,096	81
1968	744	396	484	330	16,384	5,377	68
1969	749	424	533	364	18,050	5,002	66
1970	766	425	579	392	19,566	4,412	59
1971	739	420	608	413	20,563	4,050	53
1972	—	—	615	423	20,864	3,610	46
1973	—	—	619	433	21,089	3,519	47
1974	—	—	603	423	20,559	3,429	63
1975	—	—	—	—	—	—	63

Sources: Land prices are from Sung-hwan Ban, *Hanguk Nongeop ui Seongchang* [Growth of Korean Agriculture], Seoul, 1974 (in Korean), p. 54. Wages are from National Association of Agricultural Cooperatives, *Nongchon Mulka Chongram* [Digest of Farm Village Prices], 1975, (Mimeograph, in Korean), pp. 282, 283. Sprayers, *ibid.*, pp. 256-257. Fertilizer is from Economic Planning Board, *Korea Statistical Yearbook, 1976*, p. 229. All prices are deflated to 1970 equivalents by a farm consumer price index (see Appendix C).

Note: [a] The average wage bill per farm person, based on the following 1970 per-household information: farm persons: 5.9, male labor days per year: 150.7, female labor days per year: 72.9. Taken from Ministry of Agriculture and Fisheries, *Farm Household Economy Survey 1975*, Seoul, 1976.

destabilizing impact of the collinearity between variables. Land prices in Table VIII-8 show that the price of paddy is roughly double that of upland, and the linear results from the county sample (and to a certain degree that from the provincial data) confirm this relative relationship between the productivity levels of the two most basic land categories. In addition, the value of land's contribution is of a similar order of magnitude to that of the price of land, although smaller by a factor of roughly four. This is also reasonable since land hardly pays for itself in the course of one year. The ratio of price to annual productivity would undoubtedly be higher if the value of irrigation and other required investment in land were added to its price.

If the results for land suggest some correspondence between productivity and price, those for labor present an unmistakable relationship. The cost of labor is reported in terms of a daily wage, and for our purposes this requires a reinterpretation in terms of the annual wage bill per farm person. Such a calculation is straightforward when use is made of the household size and labor contribution information mentioned earlier in the chapter. The calculations have been made and recorded in the "farm person" column of Table VIII-8. All four equations provide estimates of the annual labor contribution corresponding to one farm person which range between 9,000 and 18,000 won. In contrast to this estimated contribution of a farm person's labor, Table VIII-8's figures assign average annual wage remuneration per farm person values ranging between 12,000 and 21,000 won for the years 1959 to 1974. It is entirely reasonable that the level of reported remuneration is in general higher than our estimates of labor productivity. This is because it is very likely that much of the work done at different parts of the year is of lower productivity than the work for which wage laborers are hired. The same Household Economy Survey sources cited in Table VIII-8 indicate that only 16.8 percent of total farm household labor is hired later, the rest being family labor (75.4 percent) and exchange labor (7.8 percent).

The estimates of the marginal productivity of fertilizer from three of the four equations are also quite reasonable in comparison to the price of nitrogen fertilizer reported in Table VIII-8, and show, in fact, that the application of chemical fertilizer is quite profitable.

Similar results for pesticide and pesticide sprayers are more difficult to obtain because of the absence of a price series for pesticides and because of the difficulty of comparing the annual marginal productivity of a sprayer to its cost. Although we lack direct price series for pesticides, an estimate for any year can be obtained from survey data on average household cash expenditure on pesticides and administrative data on average per-household consumption of pesticides. The combination of these two pieces of information for 1974 yields a price per ton of 116,000 won after deflation to 1970 price levels. Although the provincial results for the marginal product of pesticides are not stable from one specification to the other because of the collinearity between fertilizer and pesticides, the two values obtained are both reasonable and suggest that pesticides are also profitable.

In sum, examination of the overall explanatory power of the equations and comparison of prices and marginal product estimates for individual variables show that the two sets of data and their summary in the form of production function estimates provide reasonable and convincing explanations for the forces behind South Korea's farm productivity. In this context, it is of greatest interest to notice the significance and sign of the coefficients for isolation in the two county equations. The t-ratios for both the linear equation (-3.3) and the log-log (-4.4) insure significance, and the coefficient and marginal product values are very close $(-8.6$ million won per kilometer and -9.5 million won per kilometer). These figures are difficult to interpret, representing as they do the average change in a county's farm value added per kilometer distance from a single center of urban industrialization. More easily interpreted results are provided below, and in the meantime, it is important only to note and remember that isolation is very important for explaining differences in county farm product, and that its statistical significance in the log-log model is greater than that for any other variable.

This demonstrated importance of isolation for understanding South Korea's pattern of regional farm productivity confirms the conclusions drawn from the provincial time series and their historical lessons. It might be argued that these results reflect nothing more than the coincidence of geography and natural productivity in South Korea. After all, Kyonggi is rich and close to Seoul while Kangwon is poor, mountainous and isolated. South Korea taken as a whole is but a single observation relating isolation to rural poverty, and any conclusions based on a single observation cannot be entirely convincing.

Such an objection can be answered, however, if we can break the country into separate regions and test our hypothesis of isolation's

Table VIII-9. Regional Regression Results with Per-household County Data

	Paddy	Upland	FPOP	FERT	SPR	RAIN (won/mm)	ISOL (won/km)	R^2/No. of Observ.
All South Korea	204 (5.4)	105 (3.6)	5,954 (3.0)	−65.9 (−0.9)	180 (3.1)	124 (3.1)	−1,490 (−6.0)	.46/167
Kyonggi and N. & S. Chungchong	216 (4.3)	99 (1.1)	2,991 (1.3)	−81.2 (−0.7)	279 (2.7)	207 (2.5)	−1,684 (−2.5)	.66/35
N. & S. Jolla and S. Kyongsang	171 (1.8)	164 (1.6)	6,305 (1.6)	159 (−1.0)	58 (0.7)	231 (3.1)	−2,096 (−4.8)	.44/52
Kangwon, N. Chungchong and N. Kyongsang	42 (0.6)	−2.0 (−0.05)	5,848 (2.1)	−122 (−1.2)	261 (3.0)	−42 (−0.6)	−909 (−2.7)	.45/44
Kyonggi, S. Chungchong N. & S. Jolla	287 (6.7)	178 (3.6)	4,710 (1.6)	35.0 (0.3)	146 (2.1)	151 (2.8)	−1,577 (−4.8)	.65/65

Note: For units, see Table VIII-3 unless otherwise noted.

importance on each individual region. In doing so below, we will also alter the data to allow for estimates of isolation's contribution which are easier to interpret than those obtained above. This is done by dividing all variables but rainfall and isolation by the county's total number of farm households. In this way, rather than dealing with total paddy and total farm population, the regression will be testing the relationships between output per household, land per household, persons per household, fertilizer per household and so forth. The results of regressions using transformed data for all of South Korea and for four sub-regions are given in Table VIII-9.

It is of course immediately apparent that the overall explanatory value of the model has dropped. This is reasonable when it is remembered that division by county household levels removes much of the coincident variation in dependent and independent variables, leaving random and unexplored portions as a larger share of the now much reduced total variation. Where the data render them significant, it is however interesting that estimates for the marginal productivity of land, fertilizer and sprayers are quite consistent with those for the untransformed data. Exceptions in this regard, however, are provided by estimates of the contribution of farm population, which are all markedly lower (by roughly a half) than the estimates above. This is, however, reasonable if we recall that farm population in the south probably was assigned by the regression analysis responsibility for contributions actually made by the more propitious climate there. Division by county household levels washes out much of this correlation, since the greater density of the south's population is also reflected in a greater household density. For this reason, the results in Table VIII-9 probably represent more accurate estimates of the average person's labor contribution than those obtained from untransformed data.

An additional advantage of per-household data is the ease of interpretation for rainfall and isolation coefficients. Rather than the marginal product per county, the coefficients now represent the marginal contribution per household. In this way we see that every millimeter of rainfall adds roughly 150 to 200 won to the average family's income. By the same token, an additional kilometer of isolation costs the average household between 1000 and 2000 won (roughly $2.50 and $5.00 respectively). These results are both large in size and statistically significant, even for sub-regions of the South Korean economy. In fact the consistently significant *t*-ratios are proof that the effect of isolation on rural productivity is not just a quirk of South Korean geography, but rather that the influence of isolation fron urban centers on farm output permeates all of rural Korea. For this reason the experience of Korea presented in this study very likely holds lessons for rural development in many parts of the world.

One final note about our analysis with altered data is needed. It has been noted above that because of the enormous variation in size of the provinces and counties, heteroskedasticity is likely to be a problem and probably introduced bias in our estimates of the standard deviations of each coefficient. Division by households in effect solves this problem and removes the source of such errors.

All in all, the regional regressions with per-household data reported in Table VIII-9 greatly enrich our results and confirm beyond any doubt that rural isolation must be an important cause for concern for anybody working on rural development. This chapter has used production function relationships to organize and summarize data from both county and provincial data, and the results, in addition to confirming the importance of traditional inputs and the correspondence of their marginal productivity with their price, have provided a supporting framework for testing the statistical importance of the isolation measure developed in the preceding chapter.

We have offered no interpretation of isolation and of its role for the rural economy. The identification of the mechanism or mechanisms through which access to cities provides benefits is a task as great as if not greater than the initial demonstration of isolation's impact. Indeed, the mechanism might involve phenomena beyond the grasp of the economist's analysis. Whether that is the case or not, practical persons everywhere can find here a basis for stressing channels of communication and transportation in economic development to a degree not warranted by the usual "infrastructure" considerations or by narrow cost-benefit calculations.

CHAPTER IX

CONCLUSION: ISOLATION AND KOREAN RURAL DEVELOPMENT

The present study has followed South Korea's farm economy over much of the hundred years between the country's opening to Japan in 1876 and the end of the study period in 1975. The regional pattern of the farm economy's growth has favored Kyonggi and certain upland dry-field provinces, although growth in the per-person and per-hectare level of output has been considerable in all regions. The strongest lesson taught to us by the massive collection of data summarized in the preceding chapters is that contact with urban industrial areas provides a very important stimulus for rural development in South Korea and accounts for a considerable portion of the regional variation throughout much of the twentieth century.

We have seen how the Korean economy in general and its agricultural component in particular have experienced phenomenal change since the days of the declining and ineffectual Yi Dynasty. Although agriculture's record seems tame in comparison to the country's industrial accomplishments, both farm output and welfare in farm communities have shown considerable growth and improvement. This experience, however, has not been shared equally by the various regions of South Korea. Although there was rough equality in output per farm at the beginning of the century, by the 1930's the region in the Northwest around Seoul had well outdistanced the other farming areas. This superior position and an equally inferior one for Kangwon Province continued to characterize the regional pattern of farm community income throughout the post-World War II years, but other changes during the same period permitted previously less affluent provinces in upland regions to out-perform the traditional rice belt areas.

In neither case, that of Kyonggi in the 1930's or the upland provinces in the 1960's and 1970's, do coincident changes in the use of traditional farm inputs point to any clear-cut reasons for the superior performances. There is no evidence that regional differences in fertilizer or pesticide use, in farmland endowment, in irrigation, in prices or even in land tenure and ownership satisfactorily explain the patterns which appeared. It is true that population densities and movements have been important, but only in that population movements were unable to keep up with the much more rapid alteration of productive conditions responsible for the balance of earlier decades.

A factor which seems most important, however, in both an historical and a statistical sense, is the influence and stimulation of urban-industrial centers in and linked with those regions which have done well. Both Seoul's location in Kyonggi and its connection by paved highway to other ascendant regions seem to provide explanations for some of the most noticeable regional asymmetries. In addition, a careful and detailed study of isolation at the county level for 1970 has shown without question that a farm community's distance from industrial cities has a very significant influence on its productivity and hence income. Rural isolation is clearly linked to rural poverty in South Korea, and this lesson has probable applications in other developing countries as well. Traditional emphasis on modern inputs and new techniques is certainly well-placed, but a catalyst which may make the new combinations work better and more quickly is the thus-far poorly understood but nevertheless clearly important touch of city life.

The effect of urban influence is poorly understood because neither this study nor those preceding it have succeeded in isolating the mechanism or mechanisms responsible. It is of course certain that easier access to urban markets makes a tremendous difference, and by itself would probably account for increases in output large enough to decidedly favor such well-located communities. Vegetable sales are the best example of this phenomenon, and both Kyonggi and the upland provinces have gained advantags in this respect. The improvement in output markets, however, is not the only and perhaps not the most important factor involved. The gains made in rice production in less isolated regions attest to this fact, because rice stores and transports well and provides no advantages for farms with better geographical market ties.

A second factor which is of obvious relevance is the quality of the non-farm labor market in and around urban centers. In the case of Kyonggi we have seen that a much slower-than-average farm population growth helped maintain superior per-household and per-person output levels, and it is clear that easier access to alternative jobs greatly aids in the absorption of a farm community's peripheral working members. Markets for other factors of production, in particular capital, have been mentioned by other writers as possible reasons for greater productivity increases, but there has been no direct proof that this is the case and that it is decisive for rural differentiation by region.

The thesis that factor markets perform better closer to urban centers rests in turn on the thesis that isolation presents impediments to factor availability in outlying areas. For traditional economic analysis this amounts to saying that the prices of modern factor inputs are higher in isolated areas and therefore more difficult to obtain. In other words, isolation impedes factor price equalization between regions and hence impedes factor use. If we go beyond the limits of traditional economic theory, however, to a world in which "cultural differences" can also play a role, the possible mechanism for urban influence becomes more complex. T. W. Schultz, writing in the early 1950's, suggests that this consideration may in fact be important:

> When factor-price equalization is based upon given wants, the cultural differences under consideration are taken as attributes of the existing pattern of wants. When the problem is approached in this way, the cultural differences between a community that has been by-passed by economic growth and progress and a community located at or near the centers of industrialization are not impediments to factor-price equalization but a part of the existing wants of the people in the two communities. It follows from this formulation that the two communities may be in equilibrium in terms of resource allocation, although great differences in the level of living exist. Another approach, the one on which this analysis rests, proceeds on the assumption that wants are not given and constant but that they are the result of cultural developments which are not independent of industrialization. One may view the changes in wants that emerge as industrialization proceeds as a movement away from a pre-industrial pattern of wants toward new, more dominant industrial-urban patterns and that the differences in wants are the result of lags in this adjustment. It is better, however, in order to simplify that analytical

problem, to introduce a value-judgement explicitly in this connection. This value-judgement is simply to the effect that the wants that characterize the communities that have been by-passed by industrial growth and progress are inferior to the wants which are emerging in the mainstream of industrialization. Given this valuation, it follows that the cultural factors that isolate the backward community and press upon it the relatively inferior wants operate as cultural impediments and, as such, impede factor-price equalization.[1]

In other words, an important part of the mechanism linking urban access to rural development stems from the fact that not only are price configurations different, but people are different also as a result of ties to the city. If it is admitted that differences in tastes and preferences for both leisure and work in different communities can influence differences in the productivity of those communities, then it is important to consider the degree to which rural contact with urban areas by itself changes the "wants" of people in those communities.

It is clear that this study has been unable to deal with this or any of the other mechanisms potentially responsible for urban influence in rural areas. Consideration of their potential, however, should only strengthen our conviction that the regional patterns of South Korea's farm development revealed by the present study have considerable significance and that those concerned with rural development in any economy may perhaps find here in the experience of South Korea lessons of general and lasting value.

---

[1] Theodore W. Schultz, "Reflections on Poverty within Agriculture," *Journal of Political Economy,* Vol. LVIII, No. 1, February 1950, pp. 1-15, reprinted in American Economic Association, *Readings in the Economics of Agriculture,* Homewood, Ill.: Richard D. Irwin, 1969, p. 336.

APPENDICES

APPENDIX A

SOUTH KOREAN PROVINCIAL STATISTICS: 1910-1940

This appendix draws from the data used in the text to analyze South Korea's farm economy in the Japanese colonial period. Its tables are based on two principal sets of statistics: annual output statistics by province for individual crops and the annual number of farm households by province, based on colonial administrative data. The prices used to combine quantity output into value output data are from Appendix C.

The crop statistics were compiled from Chosun Government General administrative reportings as published annually in its statistical yearbook. The original compilation was done at the Economic Research Center, Hitotsubashi University by Professor Shigeru Ishikawa. Household data were compiled by the author from available issues of the same Chosun Government General Statistical Yearbook.

Problems faced in the preparation of the data were of two kinds: (a) the conversion of output data from traditional Japanese/Korean units to metric tons; (b) estimation of the number of farm households for years in which data were unavailable. Solutions to the first kind of problem were straightforward, but those for the second leave open the possibility of significant error.

For years in which household data were lacking, the average annual growth rate of the number of farm households was calculated for each province between years for which data were available. These growth rates were then used to calculate estimates for intervening years. The assumption made, of course, is that actual annual rates of change did not significantly vary from the averages. This is defensible if there were not high rates of migration between provinces and to and from Korea. It is known, however, that such migration was at times significant in the colonial period, though annual data are totally lacking. It is only hoped that the years for which actual administrative data are available are numerous enough and sufficiently spaced to adequately reduce this source of error. In any event, variation in farm output levels is much greater than that in the number of farm households, and hence it is likely that the estimates presented in Table A-1 are adequate for our purposes: the comparison of per-household product for individual provinces and years.

The tables which follow are largely self-explanatory, and the details of their preparation can be found in the notes following Table A-8.

(Households)

Table A-1. Farm Households, 1910-40

Year	Seoul	Pusan	Kyonggi	Kangwon	North Chung-chong	South Chung-chong	North Jolla	South Jolla	North Kyong-sang	South Kyong-sang	Jeju	South Korea	Year
1910	0	0	201874	140476	103679	172915	168893	245000	280538	249607	0	1562982	1910
1911	0	0	220733	151942	114335	168442	174642	286376	299846	254896	0	1671212	1911
1912	0	0	226722	159389	119167	174438	188873	297610	306105	261323	0	1733627	1912
1913	0	0	232874	167201	124205	180649	204266	309285	312496	267913	0	1798889	1913
1914	0	0	240221	173440	126410	180076	200031	312634	316127	267625	0	1816564	1914
1915	0	0	247800	179914	128656	179505	195885	316021	319802	267338	0	1834921	1915
1916	0	0	246924	181652	129003	179348	197196	318777	320921	268935	0	1842756	1916
1917	0	0	246051	183407	129351	179192	198516	321557	322044	270542	0	1850660	1917
1918	0	0	245181	185179	129700	179036	199845	324361	323171	272158	0	1858631	1918
1919	0	0	244315	186968	130050	178880	201183	327190	324302	273784	0	1866672	1919
1920	0	0	243452	188775	130401	178724	202530	330043	325437	275420	0	1874782	1920
1921	0	0	242592	190599	130753	178568	203886	332921	326576	277066	0	1882961	1921
1922	0	0	241735	192441	131106	178412	205251	335824	327719	278721	0	1891209	1922
1923	0	0	240881	194301	131459	178256	206625	338753	328866	280386	0	1899527	1923
1924	0	0	240030	196179	131813	178100	208008	341707	330017	282061	0	1907915	1924
1925	0	0	239182	198075	132168	177945	209400	344687	331172	283746	0	1916375	1925
1926	0	0	238341	199995	132529	177795	210807	347701	332334	285449	0	1924951	1926
1927	0	0	238243	205393	133864	182203	213674	351997	336057	286611	0	1948042	1927
1928	0	0	238145	210937	135212	186720	216580	356346	339822	287778	0	1971540	1928
1929	0	0	238047	216630	136574	191349	219526	360749	343629	288950	0	1995454	1929
1930	0	0	237950	222477	137949	196093	222512	365207	347479	290127	0	2019794	1930
1931	0	0	237853	228482	139338	200954	225538	369720	351372	291309	0	2044566	1931
1932	0	0	237756	234649	140741	205936	228606	374288	355309	292495	0	2069780	1932
1933	0	0	237662	240985	142162	211044	231718	378918	359293	293690	0	2095472	1933
1934	0	0	240810	244179	144234	215388	236880	388338	362715	296201	0	2128745	1934
1935	0	0	244001	247416	146338	219823	242158	397994	366171	298734	0	2162635	1935
1936	0	0	244302	247504	144234	221590	239840	399574	362272	298237	0	2157553	1936
1937	0	0	244605	247593	142162	223372	237546	401161	358415	297742	0	2152596	1937
1938	0	0	245675	245392	140454	225427	239530	407096	355386	298316	0	2157276	1938
1939	0	0	246749	243211	138767	227500	241531	413119	352382	298891	0	2162150	1939
1940	0	0	247828	241049	137100	229593	243548	419231	349404	299467	0	2167220	1940

Sources: See notes following Table A-8.

Table A-2. Crop Coverage, 1910-40

(Number of Crops)

Year	Rice	Barleys	Misc. Grains	Pulses	Pota-toes	Special Crops	Fruits	Vege-tables	Live-stock	All Crops	Year
1910	1	3	6	2	0	5	0	0	0	17	1910
1911	1	3	6	3	2	5	0	0	0	20	1911
1912	1	3	6	3	2	6	3	2	0	26	1912
1913	1	3	7	3	2	6	3	2	0	27	1913
1914	1	3	7	3	2	6	3	2	0	27	1914
1915	1	3	7	4	2	6	3	2	0	28	1915
1916	1	3	7	5	2	7	3	2	0	30	1916
1917	1	3	7	5	2	7	3	2	0	30	1917
1918	1	3	7	5	2	7	3	2	0	30	1918
1919	1	3	7	5	2	7	3	2	0	30	1919
1920	1	3	7	5	2	7	3	2	0	30	1920
1921	1	3	7	5	2	7	3	2	0	30	1921
1922	1	3	7	5	2	7	3	2	0	30	1922
1923	1	3	7	5	2	7	3	2	0	30	1923
1924	1	3	7	5	2	7	3	2	0	30	1924
1925	1	3	7	5	2	7	3	2	0	30	1925
1926	1	3	7	5	2	7	3	2	0	30	1926
1927	1	3	7	8	2	8	3	2	0	34	1927
1928	1	3	7	8	2	8	3	2	0	34	1928
1929	1	3	7	8	2	8	3	2	0	34	1929
1930	1	3	7	5	2	8	3	2	0	31	1930
1931	1	3	7	5	2	8	3	12	0	35	1931
1932	1	3	7	8	2	11	5	12	0	57	1932
1933	1	3	7	8	2	11	5	12	0	57	1933
1934	1	3	7	8	2	11	5	12	0	57	1934
1935	1	3	7	8	2	11	5	12	0	57	1935
1936	1	4	7	8	2	12	5	12	0	59	1936
1937	1	4	7	8	2	12	5	12	0	59	1937
1938	1	4	7	8	2	11	5	12	0	58	1938
1939	1	4	7	8	2	11	5	12	0	58	1939
1940	1	4	7	8	2	12	5	12	0	55	1940

Sources: See Notes Following Table A-8.

Table A-3. Total Rice Output, 1910-40

(M/T)

Year	Seoul	Pusan	Kyonggi	Kangwon	North Chung-chong	South Chung-chong	North Jolla	South Jolla	North Kyong-sang	South Kyong-sang	Jeju	South Korea	Year
1910	0	0	177747	19425	34434	113188	112524	158856	204931	150494	0	971599	1910
1911	0	0	201667	44479	57055	135264	175439	199210	217506	163316	0	1193936	1911
1912	0	0	151570	46362	48794	131339	132703	168725	176233	182555	0	1038281	1912
1913	0	0	179767	52879	57356	156388	182431	195149	188123	174361	0	1186454	1913
1914	0	0	217823	63509	81686	202096	209488	228570	224480	233873	0	1461525	1914
1915	0	0	229635	60855	83346	163540	180451	223921	181780	209917	0	1333445	1915
1916	0	0	238614	58825	84072	225052	188386	233346	227520	227583	0	1483398	1916
1917	0	0	227627	61948	86635	216619	208778	227747	215204	241687	0	1486245	1917
1918	0	0	260093	89166	100217	242507	225879	303134	322018	271501	0	1814515	1918
1919	0	0	157491	88820	79277	166615	202570	318289	315418	290138	0	1618618	1919
1920	0	0	264984	100153	101354	242625	204843	269740	328684	245514	0	1757897	1920
1921	0	0	239269	90247	93904	230081	237958	302387	289141	254859	0	1737846	1921
1922	0	0	223127	103748	99710	226427	239562	318263	336849	266658	0	1814344	1922
1923	0	0	233042	94720	98371	214953	260953	326802	326257	288862	0	1843960	1923
1924	0	0	211006	101258	83249	176287	177809	289395	241765	274060	0	1554829	1924
1925	0	0	227187	101780	93539	207455	260820	324315	295257	231024	0	1741377	1925
1926	0	0	224045	104660	88261	209132	254023	331477	318697	296061	0	1826356	1926
1927	0	0	288549	123863	104101	247765	291821	349969	295907	317403	0	2019378	1927
1928	0	0	167162	108741	78025	182849	197350	329002	195324	280909	0	1539362	1928
1929	0	0	228102	119443	76573	187246	221193	266008	197669	189561	0	1485795	1929
1930	0	0	289895	139477	106953	229666	310782	367403	387631	331651	0	2163458	1930
1931	0	0	252134	116376	95344	197994	262537	288522	299445	264642	0	1776994	1931
1932	0	0	279487	127951	97688	195508	240530	354657	246081	257236	0	1799138	1932
1933	0	0	323242	138939	110342	224830	281931	325242	335882	271038	0	2011446	1933
1934	0	0	314861	120511	96749	218852	285826	322016	250441	228465	0	1837721	1934
1935	0	0	320191	144407	95332	219955	229842	255990	296590	341361	0	1903668	1935
1936	0	0	288970	134629	112609	316636	250286	338742	347650	290599	0	2080121	1936
1937	0	0	473337	193518	170379	379434	402285	427316	415815	420871	0	2882955	1937
1938	0	0	428241	186328	155274	333099	378148	374224	409317	342210	0	2606841	1938
1939	0	0	205337	163160	53439	105035	101509	225653	107066	184153	0	1145352	1939
1940	0	0	270226	130022	110660	254669	352413	432850	386192	346810	0	2283842	1940

Sources: See notes following Table A-8.

Table A-4. Farm Output (1970 Prices), 1910-40

(Million 1970 Won)

Year	Seoul	Pusan	Kyonggi	Kangwon	North Chung-chong	South Chung-chong	North Jolla	South Jolla	North Kyong-sang	South Kyong-sang	Jeju	South Korea	Year
1910	0	0	16602	4451	4425	10738	10197	16620	22129	15136	0	100303	1910
1911	0	0	20877	7939	6611	13121	15621	20643	25321	18125	0	128262	1911
1912	0	0	20277	10555	6960	14404	13520	19849	22446	20298	0	128312	1912
1913	0	0	22030	10937	7373	15333	17407	23747	23468	21031	0	141330	1913
1914	0	0	25695	12751	10445	20361	20787	26994	26870	25203	0	169111	1914
1915	0	0	29117	12724	11171	18749	18252	28636	24236	25088	0	167978	1915
1916	0	0	29965	12311	11155	23073	18797	27991	27128	26115	0	176537	1916
1917	0	0	29569	13444	12009	22424	19990	30405	29390	28951	0	186185	1917
1918	0	0	34975	20403	15663	26396	23478	38212	44755	33291	0	237176	1918
1919	0	0	23521	17468	13429	19164	21856	41454	45600	37368	0	219863	1919
1920	0	0	36704	20069	16295	26999	22564	39091	45794	33386	0	240906	1920
1921	0	0	34595	18874	15663	26190	24730	40426	42197	34140	0	236819	1921
1922	0	0	31809	19747	15015	25423	24396	42920	46672	34759	0	240743	1922
1923	0	0	32042	18959	15111	25015	26369	42585	43548	34670	0	238301	1923
1924	0	0	27808	18334	13690	21385	19814	42050	37608	35511	0	216203	1924
1925	0	0	30985	19688	15316	25402	27290	44437	42617	31720	0	237459	1925
1926	0	0	30419	20049	15193	25861	26792	45256	44318	36482	0	244372	1926
1927	0	0	36133	22608	17539	29857	29989	46957	42741	37535	0	263363	1927
1928	0	0	24424	20475	14378	22970	22011	46451	31431	34056	0	216200	1928
1929	0	0	30781	21999	14940	24010	24074	41137	31938	26510	0	215392	1929
1930	0	0	35135	23944	16792	27588	31187	49746	49333	38749	0	272478	1930
1931	0	0	32813	21072	15796	24610	27079	38753	42276	33508	0	235911	1931
1932	0	0	45708	26983	19366	29835	29433	50266	40787	37261	0	279643	1932
1933	0	0	49727	27594	19612	32256	33042	47172	47100	38037	0	294543	1933
1934	0	0	51241	23608	18976	31752	33841	48742	40036	34953	0	283153	1934
1935	0	0	53827	27437	20160	33426	31318	47538	46303	47021	0	307034	1935
1936	0	0	50394	24276	20362	39142	31799	46657	45070	35606	0	293309	1936
1937	0	0	69859	33355	27493	48080	46308	63716	57353	53513	0	399681	1937
1938	0	0	66237	31688	25976	42771	42375	54067	55708	44020	0	362846	1938
1939	0	0	39885	25711	14671	20949	19417	47329	28871	33175	0	230010	1939
1940	0	0	40998	24629	19678	33228	40227	58663	53043	44947	0	315417	1940

Sources: See notes following Table A-8.

Table A-5. Per-Household Farm Output, 1910-40

(thousand 1970 won)

Year	Seoul	Pusan	Kyonggi	Kangwon	North Chung-chong	South Chung-chong	North Jolla	South Jolla	North Kyong-sang	South Kyong-sang	Jeju	South Korea	Year
1910	0	0	82	31	42	62	60	67	78	60	0	64	1910
1911	0	0	94	52	57	77	89	72	84	71	0	76	1911
1912	0	0	89	66	58	82	71	66	73	77	0	74	1912
1913	0	0	94	65	59	84	85	76	75	78	0	78	1913
1914	0	0	106	73	82	113	103	86	84	94	0	93	1914
1915	0	0	117	70	86	104	93	90	75	93	0	91	1915
1916	0	0	121	67	86	128	95	87	84	97	0	95	1916
1917	0	0	120	73	92	125	100	94	91	107	0	100	1917
1918	0	0	142	110	120	147	117	117	138	122	0	127	1918
1919	0	0	96	93	103	107	108	126	140	136	0	117	1919
1920	0	0	150	106	124	151	111	118	140	121	0	128	1920
1921	0	0	142	99	119	146	121	121	129	123	0	125	1921
1922	0	0	131	102	114	142	118	127	142	124	0	127	1922
1923	0	0	133	97	114	140	127	125	132	123	0	125	1923
1924	0	0	115	93	103	120	95	123	113	125	0	113	1924
1925	0	0	129	99	115	142	130	128	128	111	0	123	1925
1926	0	0	127	100	114	145	127	130	133	127	0	126	1926
1927	0	0	151	110	131	163	140	133	127	130	0	135	1927
1928	0	0	102	97	106	123	101	130	92	118	0	109	1928
1929	0	0	129	101	109	125	109	114	92	91	0	107	1929
1930	0	0	147	107	121	140	140	136	141	133	0	134	1930
1931	0	0	137	92	113	122	120	104	120	115	0	115	1931
1932	0	0	192	114	137	144	128	134	114	127	0	135	1932
1933	0	0	209	114	137	152	142	124	131	129	0	140	1433
1934	0	0	212	96	131	147	142	125	110	118	0	133	1934
1935	0	0	220	110	137	152	129	119	126	157	0	141	1935
1936	0	0	206	98	141	176	132	116	124	119	0	135	1936
1937	0	0	285	134	193	215	194	158	160	179	0	185	1937
1938	0	0	269	129	184	189	176	132	156	147	0	168	1938
1939	0	0	161	105	105	92	80	114	81	110	0	106	1939
1940	0	0	165	102	143	144	165	139	151	150	0	145	1940

Sources: See notes following Table A-8.

Table A-6. Per-Household Rice Output, 1910–40

(1970 won)

Year	Seoul	Pusan	Kyonggi	Kangwon	North Chung-chong	South Chung-chong	North Jolla	South Jolla	North Kyong-sang	South Kyong-sang	Jeju	South Korea	Year
1910	0	0	67348	10577	25403	50069	50961	49595	55875	46117	0	47548	1910
1911	0	0	69883	22391	38169	61423	76839	53208	55485	49008	0	54645	1911
1912	0	0	51135	22248	31319	57591	53742	43364	44037	53434	0	45810	1912
1913	0	0	59046	24190	35321	66217	68313	48262	46047	49780	0	50448	1913
1914	0	0	69358	28008	49427	85843	80106	55922	54315	66843	0	61540	1914
1915	0	0	70882	25872	49551	69687	70463	54198	43477	60060	0	55585	1915
1916	0	0	73915	24770	49848	95982	73072	55990	54228	64728	0	61573	1916
1917	0	0	70762	25835	51230	92466	80444	54175	51113	68331	0	61428	1917
1918	0	0	81142	36830	59102	103606	86454	71484	76217	76305	0	74674	1918
1919	0	0	49307	36336	46627	71245	77017	74409	74394	81058	0	66325	1919
1920	0	0	83255	40581	59451	103838	77363	62514	77253	68184	0	71721	1920
1921	0	0	75442	36217	54933	98555	89272	69474	67722	70359	0	70595	1921
1922	0	0	70602	41236	58172	97075	89276	72490	78620	73179	0	73381	1922
1923	0	0	74000	37288	57237	92236	96601	73791	75883	78802	0	74252	1923
1924	0	0	67240	39480	48308	75711	65385	64780	56035	74320	0	62334	1924
1925	0	0	72654	39304	54134	89174	95272	71969	68194	62277	0	69505	1925
1926	0	0	71902	40028	50940	89971	92170	72920	73351	79333	0	72572	1926
1927	0	0	92641	46127	59483	104013	104464	76049	67351	84707	0	79290	1927
1928	0	0	53690	39431	44139	74904	69698	70620	43965	74664	0	59722	1928
1929	0	0	73294	42174	42885	74849	77070	56401	44000	50180	0	56953	1929
1930	0	0	93187	47953	59303	89585	106833	76949	85328	87437	0	81930	1930
1931	0	0	81082	38959	52339	75363	89038	59691	65186	69487	0	66479	1931
1932	0	0	89915	41708	53091	72616	80479	72478	52975	67269	0	66488	1932
1933	0	0	104033	44100	59369	81486	93065	65654	71506	70590	0	73422	1933
1934	0	0	100011	37750	51307	77720	92294	63426	52813	58998	0	66032	1934
1935	0	0	100374	44644	49829	76535	72599	49198	61955	87404	0	67330	1935
1936	0	0	90475	41606	59718	109298	79821	64844	73402	74531	0	73744	1936
1937	0	0	148016	59784	91672	129930	129536	81477	88739	108121	0	102442	1937
1938	0	0	133331	58079	84560	113024	120755	70313	88097	87744	0	92430	1938
1939	0	0	63652	51313	29456	35314	32146	41780	23240	47127	0	40518	1939
1940	0	0	83402	41258	61738	84844	110680	78974	84543	88582	0	80606	1940

Sources: See notes following Table A-8.

Table A-7. *Per-Household Barleys Output, 1910-40*

(1970 won)

Year	Seoul	Pusan	Kyonggi	Kangwon	North Chung-chong	South Chung-chong	North Jolla	South Jolla	North Kyong-sang	South Kyong-sang	Jeju	South Korea	Year
1910	0	0	2965	2212	7235	4385	3374	11442	11294	6845	0	6825	1910
1911	0	0	5630	3573	7667	7019	4545	11177	11637	12725	0	8719	1911
1912	0	0	8670	3832	10126	8227	5451	10359	12130	11685	0	9285	1912
1913	0	0	9288	4135	10819	9472	6398	14304	11671	15826	0	10855	1913
1914	0	0	7527	4655	11508	8992	7476	11548	12947	10995	0	9816	1914
1915	0	0	8366	4756	12120	10266	7877	12836	12875	14235	0	10820	1015
1916	0	0	9146	5719	12882	8359	6110	10325	11447	12113	0	9706	1916
1917	0	0	8496	4939	11493	9374	6466	12521	13958	14759	0	10785	1917
1918	0	0	9670	4944	11911	12510	7335	12775	15182	15041	0	11665	1918
1919	0	0	11130	8078	18247	12489	7404	15114	22660	20051	0	15059	1919
1920	0	0	11939	7708	16988	12503	8331	15504	19679	20638	0	14777	1920
1921	0	0	11225	6727	17794	13170	9420	18073	20557	21203	0	15512	1921
1922	0	0	9463	7729	14728	10715	7481	15653	19270	19321	0	13806	1922
1923	0	0	9416	7258	13496	11435	7018	14060	16965	14752	0	12329	1923
1924	0	0	10430	8176	14630	11682	8284	16610	20961	19194	0	14595	1924
1925	0	0	11270	7896	16122	12651	9984	17944	21795	20489	0	15628	1925
1926	0	0	10733	7955	15728	12056	9145	16453	20448	17003	0	14377	1926
1927	0	0	9037	7366	15134	11915	9374	16504	18483	16154	0	13612	1927
1928	0	0	7692	5523	11626	10062	9124	17543	16660	18264	0	12981	1928
1929	0	0	9350	6914	12416	10604	9573	16823	19338	18687	0	13863	1929
1930	0	0	10633	6823	13311	12332	10908	17775	18885	18363	0	14413	1930
1931	0	0	12103	6522	14502	12313	9825	17882	19376	20021	0	14835	1931
1932	0	0	12773	5655	14983	12221	12013	17827	19661	20832	0	15213	1932
1933	0	0	12524	5579	12074	11003	11785	17654	18423	20891	0	14572	1933
1934	0	0	14069	5776	14199	12414	13655	20313	18722	21803	0	15921	1934
1935	0	0	14843	6916	13967	12810	12770	22891	20560	25346	0	17338	1935
1936	0	0	9183	4166	12935	11553	12262	22625	17329	20613	0	14881	1936
1937	0	0	17273	9491	24783	20828	17907	25592	27026	26758	0	21799	1937
1938	0	0	11876	6556	19059	16135	15038	17578	24172	18671	0	16576	1938
1939	0	0	17332	7317	22706	17808	15142	23414	21953	21771	0	18885	1939
1940	0	0	10848	3871	21103	18677	19540	23535	26008	19786	0	18660	1940

Sources: See notes following Table A-8.

Table A-8. Per-Household Vegetable Output, 1910-40

(1970 won)

Year	Seoul	Pusan	Kyonggi	Kangwon	North Chungchong	South Chungchong	North Jolla	South Jolla	North Kyongsang	South Kyongsang	Jeju	South Korea	Year
1910	0	0	0	0	0	0	0	0	0	0	0	0	1910
1911	0	0	0	0	0	0	0	0	0	0	0	0	1911
1912	0	0	12507	11504	3126	3171	2588	2554	2354	580	0	4451	1912
1913	0	0	8268	9352	415	333	1794	2103	1982	1086	0	3073	1913
1914	0	0	11470	8973	2682	4053	2910	2553	1879	1516	0	4272	1914
1915	0	0	14560	7461	4397	4382	3034	2591	2904	2698	0	5104	1915
1916	0	0	12877	6375	2636	4041	3540	2614	2435	2567	0	4561	1916
1917	0	0	13503	7097	4553	3325	2548	2512	6406	2621	0	5346	1917
1918	0	0	13722	11199	6119	5314	4153	4875	7541	4344	0	7109	1918
1919	0	0	14792	8117	3831	4297	4773	4685	4416	4312	0	6163	1919
1920	0	0	16771	8503	6781	5484	6580	5596	4951	4309	0	7217	1920
1921	0	0	16543	8249	8874	6049	5175	3934	4579	3736	0	6756	1921
1922	0	0	17127	8424	5849	6775	5069	5248	5293	4339	0	7130	1922
1923	0	0	14126	7658	5525	7378	4162	2773	2799	1012	0	5231	1923
1924	0	0	11295	6405	5118	8033	4942	4421	5514	3891	0	6043	1924
1925	0	0	12730	7361	5767	9848	5702	4182	5376	3120	0	6428	1925
1926	0	0	13709	8205	5625	12375	5248	4166	5010	4023	0	6869	1926
1927	0	0	13784	8423	5519	13032	5566	4520	6242	4811	0	7384	1927
1928	0	0	11262	7782	4753	7646	4509	2873	5083	3508	0	5646	1928
1929	0	0	13302	7914	5063	8243	4707	4021	6483	3895	0	6508	1929
1930	0	0	12181	9100	5027	8663	4418	3364	3779	2363	0	5706	1930
1931	0	0	12218	8183	5048	8356	3039	3704	5228	3625	0	5921	1931
1932	0	0	53400	24239	27411	30493	15651	8392	10624	10163	0	20286	1932
1933	0	0	55267	21672	25051	31083	16893	10896	9382	10209	0	20468	1933
1934	0	0	62412	17376	25869	29406	16901	11143	10322	11621	0	21071	1934
1935	0	0	66539	18024	27723	29943	21027	11108	9962	14052	0	22515	1935
1936	0	0	70229	16808	24536	29319	21069	9635	8900	9448	0	21458	1936
1937	0	0	78826	20259	23310	30934	23161	12755	10440	12436	0	24428	1937
1938	0	0	85369	20819	26381	29486	19701	11147	11988	13973	0	25087	1938
1939	0	0	54801	14229	15506	21631	16553	12022	8083	14598	0	18607	1939
1940	0	0	45930	17690	16072	20929	17128	11030	12804	13783	0	18481	1940

Sources: See notes following this table.

Notes to Tables in Appendix A

Table A-1

Original data were used for the following years only: 1910, 1911, 1913, 1915, 1926, 1933, 1935, 1937 and 1941. They were taken from individual years of the Chosun Government General, *Tokei Nenbo* [Statistical Yearbook], Seoul. Data for years other than those listed above were estimated for each province by using the average rate of annual change between known years.

Table A-2

This table gives the number of output crops in each category for which crop data were available. They were tabulated from the crop data cited in the note to Tables A-3 and A-4 below.

Table A-3

These data were converted to tons from data in *seok* taken from Shigeru Ishikawa, *Chosen Nogyo Seisankaku no Shukei, Senzen no Bu* [Collected Statistics of Chosun Agricultural Production, Pre-war Part], Kunitachi, Japan: Economic Research Center, Hitotsubashi University, 1973 (mimeographed, in Japanese). A *seok* is a unit of volume, and hence its conversion factor to metric tons is different for different crops; the factor used for rice in this study is 0.144 metric tons per *seok*.

Table A-4

These data were computed using 1970 crop prices per metric ton (see Appendix C) and individual crop output levels as reported by the Chosun Government General and compiled in Ishikawa, cited above. The conversion factors from Korean/Japanese volume and weight units to metric tons were based on information in Republic of Korea, Ministry of Agriculture and Fisheries, *Yearbook of Agriculture and Forestry Statistics, 1975,* pp. 459-461. For crop coverage by category in each year, see Table A-2 above.

Table A-5

These data were computed by dividing data in Table A-4 by corresponding data from Table A-1.

Tables A-6, A-7 and A-8

Data in these tables were computed from the rice, barleys and vegetables components of Table A-4's total value computation. Each component was then divided by the corresponding household data from Table A-1. For crop coverage information, see Table A-2. The only two crops reported until 1932 in the vegetable category were Chinese cabbage and the Korean giant white radish.

APPENDIX B

SOUTH KOREAN PROVINCIAL STATISTICS: 1941-1975

The data in this appendix are drawn from the vast amount of South Korean provincial time-series data collected and compiled from tables in individual editions of the Republic of South Korea, Ministry of Agriculture and Forestry, *Yearbook of Agriculture and Forestry Statistics,* 1953-1976, Seoul, and from the 1952 edition of the same yearbook, published in Pusan, which contains provincial statistics for every year from 1936 to 1951. Corrections were made for numerous errors and inconsistencies found in the original data, and some effort was made to compensate for changes in boundaries which occurred during the time period covered. Gross output value calculations made from the crop data are also reported below. General issues dealing with data in several or all of the tables will be discussed briefly in the following passages. For a more detailed discussion of individual tables, see the notes which follow Table B-23.

The data presented here are administrative data, that is, they are collected by government agencies in the process of population registration, taxation, and government sale or purchase of goods. Crop statistics are compiled on the basis of district (*Myeon*) planted area estimates and estimated yields. Changes in the quality of this estimation process introduce difficulties and are dealt with in a subsequent paragraph. These administrative data, therefore, are of different quality from data obtained by census or sample survey, and this should be kept in mind when reviewing the conclusions they suggest. The data are probably less reliable as an indication of absolute levels, but are very likely much more dependable for comparisons between provinces, since the collection agencies are part of a highly centralized administrative system with uniform standards for personnel and procedures. For a discussion of the differences between administrative and census reporting in the all-important farm population category, see the treatment on pages 151-152, above, including Footnote 2 to Chapter VIII on page 152.

Although the published provincial data were obtained for every year in the period covered, there were scattered years in which data for one item or category of items were unreported. In such cases, a single unreported year's data were estimated by averaging the levels for the preceding and following years. In the rare case where more than one consecutive year's data were unreported, individual average annual growth rates for each province were used to estimate the missing figures. For details, see the notes following Table B-23.

Analysis of Korean agricultural statistics during a period which spans the end of World War II or the Korean War or both (as the present study does) faces the formidable problem of correcting for changes over time due solely to changes in South Korean boundaries, or more specifically, the boundaries of Kvonggi and Kangwon Provinces. The only correct procedure would be to collect county-level data for those provinces for the period before the Korean War, but these data were not available. For the present study the method used by Sung-hwan Ban in his *Hangug Nongeop ui Seongchang* (Growth of Korean Agriculture), Seoul: KDI Press, 1974 (in Korean), was used. Dr. Ban found the ratio of pre-World War II and post-Korean War average levels of planted paddy, planted upland and total planted land for both Kyonggi and Kangwon. These ratios were then used to adjust downward the crop output levels, with the upland ratio applied to all crops except rice. Farm population estimates were also calculated using the ratios for total planted area. Although these calculations are an improvement when looking at total output data over time, they introduce unnecessary distortions into the per-capita and per-household calculations most valuable for provincial comparisons. It should be noted that accurate comparisons over the whole period can be made using the provincial time series for all provinces except Kyonggi, Kangwon and South Jolla (since Jeju was considered part of South Jolla before 1945). The correction ratios used were (Kyonggi) .831, .610 and .730; (Kangwon) .548, .327 and .386 for paddy, upland and total planted land, respectively. Finally, it should also be noted that data for inter-war years, 1945-49, are given as reported in administrative statistics. For Kyonggi and Kangwon, these refer to areas different from those for both preceding and following years.

As noted above, yield estimates were combined with estimates of planted land to obtain the district-level estimates of individual crop output eventually reported in the administrative tables used for this study. Although there is undoubtedly considerable room for error in this method, if procedures remain unchanged from year to year, relative changes over time and between provinces should still be reflected in the reported data. Such was the case for South Korea between 1941 and 1975 with one great exception. Between the years 1962 and 1964 the Ministry of Agriculture and Forestry revised its methods for estimating the yields of major foodgrains after test samples showed that there was serious underestimation. This posed serious problems for the estimation of comparable crop output levels for the years before the revision. When did the apparent overestimation begin? Was it uniform from region to region?

There is no satisfactory answer to the first question, but it is likely that significant underestimation and under-reporting of output began as early as 1941, when forced collection of grain output was begun by the Japanese as part of their war effort. It is difficult to say whether this under-reporting became more or less serious during the period of American military government, during the Korean War, or during the period of the Rhee government in the 1950's. Consequently, this study has revised such data back to 1941. The correction factors used were obtained from data in the mid-1960s, when both the old and the new sampling methods were used in the same year(s). The method of correction, however, was more accurate for rice than for other crops.

The greater accuracy in this study's rice corrections is due to the availability of provincial-level correction factors for rice, while only national average correction factors are available for the other crops (barley, naked barley, wheat, and rye). The provincial-level correction factors for rice are given in Table B-15. They were obtained from hand-written records of the sample tests in the Ministry of Agriculture and Forestry archives, but the Ministry's own adjustments in retrospective statistics published after 1965 made use of only a single national average correction factor, applying that same factor to all provinces. Hence, this study's revisions of rice output statistics prior to 1965 are more accurate than those reported by the Ministry of Agriculture and Forestry and other researchers, though the difference is admittedly of limited significance. Nevertheless, it is interesting to note in Table B-15 that Kyonggi Province rice output had traditionally been more accurately reported than had output in other provinces, and hence required less correction. In other words, Kyonggi's superior output in earlier years clearly survived the more generous corrections made for other provinces.

The correction factors for barley, naked barley, wheat, and rye are 1.946, 1.324, 2.264, and 2.099, respectively. No corrections were made for white potatoes or sweet potatoes, although late in the preparation of the study it was pointed out that similar tests in 1966 showed the need for correction factors of 1.333 and 1.785 respectively.

Value output calculations were made using crop prices from Appendix C and individual provincial data for all crops in all years. Two types of calculations were used, one with constant 1970 prices and the other with current prices deflated by the National Agricultural Cooperatives Association farm consumer price index. The former method is straightforward and allows comparison of results for this period with those for the earlier period reported in Appendix A. The latter method introduces some complications because there are years when crop outputs are reported but corresponding prices are not. In such cases the constant 1970 price is used rather than the deflated current price.

Value output estimates for crop categories as well as total output were made and are reported below for rice, barleys and vegetables. The number of crops covered in different years for these and other categories is given in Table B-13.

Table B-1. Farm Population, 1941–75

(persons)

Year	Seoul	Pusan	Kyonggi	Kangwon	North Chung-chong	South Chung-chong	North Jolla	South Jolla	North Kyong-sang	South Kyong-sang	Jeju	South Korea	Year
1941	0	0	1066139	524062	804536	1361029	1364659	2249487	1995656	1664650	0	12258382	1941
1942	0	0	1068754	513091	796492	1372672	1375155	2265016	1968994	1660953	0	12232589	1942
1943	0	0	1131891	625284	884636	1454430	1508089	2415672	2096517	1797198	0	13321184	1943
1944	0	0	1161083	645850	793384	1404660	1420197	1903292	2237047	1885626	0	12905019	1944
1945	0	0	1630515	666416	702132	1354890	1332305	1390913	2377578	1974055		11428804	1945
1946	0	0	1779123	752680	769057	1394598	1423502	1801509	2382695	2183594	103733	12590494	1946
1947	0	0	1927732	838994	835983	1434307	1514700	2212106	2387812	2393133	207466	13752183	1947
1948	35313	0	1943975	761957	879368	1562865	1581420	2343660	2426965	2337932	210816	14084274	1948
1949	70627	0	1960219	684970	922753	1691424	1648140	2475215	2466119	2282731	214167	14416365	1949
1950	70629	0	1331310	664371	877909	1638093	1413648	2176121	2259599	2212866	223069	12867615	1950
1951	63913	0	1320322	645570	889026	1635767	1422013	2206965	2305652	2161027	214694	12864952	1951
1952	57197	0	1309335	626770	900144	1633441	1430379	2237809	2351706	2109188	206320	12862289	1952
1953	61205	0	1356096	628971	907958	1643018	1528403	2343613	2339985	2114956	225321	13149526	1953
1954	56604	0	1367666	637307	923258	1685998	1535743	2300160	2321697	2130726	212443	13171592	1954
1955	55807	0	1374846	720449	930353	1689745	1501493	2368113	2346196	2099681	212513	13299196	1955
1956	52546	0	1425617	733278	941251	1704159	1545934	2383387	2336162	2101260	218105	13442199	1956
1957	49232	0	1458526	741848	946594	1721001	1539136	2436438	2363384	2110735	222735	13589629	1957
1958	45968	0	1478400	750087	950125	1728857	1549480	2437316	2469187	2111004	228623	13749047	1958
1959	41388	0	1516976	777997	986512	1781646	1615149	2524088	2518951	2134865	225430	14123002	1959
1960	40906	0	1571586	814325	1012541	1860744	1638387	2609706	2580100	2198284	231617	14558196	1960
1961	40018	0	1574706	820635	1033039	1851284	1626791	2594687	2549832	2181339	234789	14507120	1961
1962	122349	58409	1556665	775557	1087953	2061793	1608918	2719765	2691938	2182107	246649	15112103	1962
1963	115812	59582	1574085	884247	1086854	1956711	1692399	2736105	2712805	2196295	251444	15266325	1963
1964	114345	59586	1611612	922095	1099744	2007552	1701013	2787842	2759583	2236960	252687	15553019	1964
1965	114235	61561	1624419	926313	1115400	2028403	1761260	2859272	2772832	2286816	261064	15811575	1965
1966	108025	60486	1603481	933987	1111159	2026311	1773007	2884092	2761963	2263082	255113	15780706	1966
1967	102986	58859	1620914	959468	1151279	2049904	1835974	2954112	2797825	2277275	269490	16078086	1967
1968	90850	55091	1593238	938719	1134178	2036180	1814819	2968221	2754764	2251027	270567	15907664	1968
1969	78622	49898	1552753	912114	1105754	1997691	1771922	2952790	2681111	2214354	271903	15588912	1969
1970	71216	46008	1530309	898587	1075365	1943075	1741000	2865489	2630723	2153901	265660	15221337	1970
1971	58609	40808	1486376	880329	1027157	1869603	1689638	2772376	2550089	2079764	257079	14711828	1971
1972	56784	38238	1488871	863186	1034371	1868828	1687623	2768571	2536920	2070461	263091	14676944	1972
1973	56503	33289	1488633	870645	1031192	1860052	1688726	2711971	2528444	2049401	265710	14644566	1973
1974	35395	31804	1423635	872060	920422	1732332	1580689	2500660	2271369	1844270	246559	13459195	1974
1975	31599	19869	1344978	727919	886664	1583398	1455010	2155983	2166989	1659317	230950	12262676	1975

Sources: See notes following Table B-23.

Table B-2. Farm Households, 1941-75

(households)

Year	Seoul	Pusan	Kyonggi	Kangwon	North Chungchong	South Chungchong	North Jolla	South Jolla	North Kyongsang	South Kyongsang	Jeju	South Korea	Year
1941	0	0	179700	87733	136437	230510	243388	410951	338933	293624	0	2127758	1941
1942	0	0	180141	85897	135073	232482	245260	413788	334405	292972	0	2123281	1942
1943	0	0	190783	104679	150021	246329	268969	441311	356063	317004	0	2312235	1943
1944	0	0	197719	97737	135959	238816	255223	427042	349390	308369	0	2238856	1944
1945	0	0	280349	90796	121897	231303	241478	412774	342718	299734	0	2021049	1945
1946	1560	0	281507	105970	130348	233910	256991	398669	350511	314088	23187	2096743	1946
1947	3120	0	282665	121143	138800	236518	272505	384564	358304	328442	46374	2172435	1947
1948	7861	0	307020	126433	148552	258571	277022	402643	389224	358740	47067	2323134	1948
1949	12602	0	331375	131723	158304	280625	281539	420722	420145	389038	47760	2473833	1949
1950	12602	0	238108	117117	148721	269457	251027	413617	388743	376054	45302	2260748	1950
1951	11542	0	234565	114698	149429	271494	249324	406930	395581	368335	45370	2247270	1951
1952	10483	0	231022	112279	150137	273532	247622	400243	402420	360616	45438	2233792	1952
1953	10724	0	229649	111906	151845	274601	257391	396315	407691	362232	46099	2248453	1953
1954	10690	0	231719	110118	151766	274181	255183	392996	397647	363223	46814	2234337	1954
1955	10161	0	233366	128444	151604	270867	243439	394055	383204	357444	44746	2217330	1955
1956	9725	0	235857	130061	149080	266560	247115	388631	378787	347967	45292	2199075	1956
1957	8972	0	239972	129512	149729	266856	247241	393707	379233	347483	46794	2209499	1957
1958	8282	0	238093	130841	148635	261012	247307	393653	393840	347775	47317	2216755	1958
1959	7890	0	244237	135873	151475	269314	250950	398944	404949	354839	47489	2265960	1959
1960	7405	0	254049	141279	159958	283173	258371	415122	416524	362654	49723	2348258	1960
1961	6837	0	254619	142367	161121	279181	252858	413779	409842	354509	50747	2325860	1961
1962	20391	9988	256367	140489	170495	314198	257949	439326	443499	362486	55141	2470329	1962
1963	19025	10015	251106	149856	167545	291821	262741	425852	428622	352623	56387	2415593	1963
1964	18813	9873	254926	156550	169886	296486	266511	432625	429637	358245	56776	2450308	1964
1965	18552	10235	259144	157334	173590	304886	275890	441470	441269	366102	58427	2506899	1965
1966	17583	10052	260624	159463	176213	310745	280622	451408	448304	368594	56666	2540274	1966
1967	16558	9581	263995	161422	180907	317886	295929	460551	450605	370933	58457	2586864	1967
1968	14380	9024	260646	158452	179984	319202	294748	467083	448111	368632	58264	2578526	1968
1969	12668	8261	255853	154629	176882	316127	291529	466075	441463	364667	58090	2546244	1969
1970	11408	7596	254942	152001	173803	312103	286699	461661	436707	359890	57720	2514534	1970
1971	9402	6791	252828	150064	169534	308409	282084	458535	431041	355571	57266	2481525	1971
1972	9185	6311	250443	144861	168814	304674	278435	454954	426213	350693	57261	2451844	1972
1973	9561	5541	250078	145303	169139	304443	279139	455102	425471	348591	57909	2450277	1973
1974	5666	5600	254574	151045	159832	299754	274054	435636	405013	334855	55171	2381200	1974
1975	6446	4934	262463	137875	162405	295295	271540	431640	416481	334243	55736	2379058	1975

Sources: See notes following Table B-23.

Table B-3. Share of Full-time Households, 1952-75

(per mil)

Year	Seoul	Pusan	Kyonggi	Kangwon	North Chung-chong	South Chung-chong	North Jolla	South Jolla	North Kyong-sang	South Kyong-sang	Jeju	South Korea	Year
1952	645	0	916	873	903	899	904	853	897	882	862	887	1952
1953	641	0	920	869	911	908	917	867	903	880	852	894	1953
1954	600	0	914	856	906	902	915	870	894	867	878	888	1954
1955	681	0	911	886	918	907	924	882	907	874	907	898	1955
1956	660	0	915	893	920	912	930	892	911	886	896	905	1956
1957	691	0	914	893	924	911	922	885	909	884	909	902	1957
1958	748	0	915	894	930	914	915	886	908	876	891	901	1958
1959	720	0	920	907	929	914	926	897	918	890	928	910	1959
1960	720	0	910	901	929	914	924	888	917	891	907	907	1960
1961	741	0	916	908	931	917	923	887	918	890	913	908	1961
1962	816	711	919	926	927	926	920	885	917	892	925	909	1962
1963	840	733	921	923	922	927	922	885	917	894	920	910	1963
1964	837	713	923	918	926	928	926	883	918	888	921	909	1964
1965	810	711	909	917	926	913	929	879	921	897	914	907	1965
1966	779	660	906	903	925	897	923	864	908	895	919	898	1966
1967	697	667	873	897	905	860	868	830	900	869	893	870	1967
1968	612	626	856	884	879	844	854	812	891	852	877	853	1968
1969	614	626	874	889	883	856	837	801	901	859	878	856	1969
1970	697	680	899	910	907	875	850	841	923	891	901	883	1970
1971	824	748	932	935	933	904	858	880	952	929	928	913	1971
1972	814	757	938	937	935	899	851	874	950	925	921	910	1972
1973	652	542	882	896	905	842	810	789	896	864	816	851	1973
1974	487	372	800	839	890	780	810	768	848	770	767	803	1974
1975	578	452	803	838	898	766	831	788	831	766	789	805	1975

Sources: See notes following Table B-23.

Table B-4. Share of Wage-labor Households, 1968-75

(per mil)

Year	Seoul	Pusan	Kyonggi	Kangwon	North Chung-chong	South Chung-chong	North Jolla	South Jolla	North Kyong-sang	South Kyong-sang	Jeju	South Korea	Year
1968	65	6	24	23	23	28	64	46	10	24	15	31	1968
1969	50	17	23	24	36	36	52	51	10	27	17	32	1969
1970	49	16	32	23	34	33	67	53	14	29	22	36	1970
1971	42	27	44	22	41	33	73	55	18	31	26	39	1971
1972	29	19	37	24	39	37	79	59	20	33	31	41	1972
1973	21	20	38	20	22	37	77	53	20	32	37	39	1973
1974	43	—	61	29	40	31	82	34	14	16	42	36	1974
1975	49	36	72	25	15	27	76	31	14	23	61	35	1975

Sources: See notes following Table B-23.

Table B-5. Securely Irrigated Paddy, 1955-75

(hectares)

Year	Seoul	Pusan	Kyonggi	Kangwon	North Chung-chong	South Chung-chong	North Jolla	South Jolla	North Kyong-sang	South Kyong-sang	Jeju	South Korea	Year
1955	1318	0	72751	23182	25945	71030	92318	86236	79884	87069	356	540089	1955
1956	1278	0	73649	24000	27666	71030	92318	86300	82439	85911	413	545004	1956
1957	1041	0	75502	24184	29410	71244	92318	107492	85628	84992	478	572289	1957
1958	1065	0	76807	25038	31660	74163	92909	107498	87088	85208	397	581833	1958
1959	1169	0	76666	26059	32332	74800	92317	107545	87369	90319	553	589129	1959
1960	1070	0	77684	25832	33968	79263	97145	109420	89246	89424	625	603677	1960
1961	956	0	80808	26345	34887	81229	101353	112598	92313	89795	588	620872	1961
1962	1086	0	84753	30199	34964	86312	105025	118141	96825	90424	620	648349	1962
1963	5617	1995	80530	31708	36014	87808	101347	120046	99439	88647	661	653812	1963
1964	4939	1400	84619	34028	39050	93517	107834	124122	103848	96175	664	690196	1964
1965	4076	1515	91939	31659	39728	94289	109571	120078	110505	98400	583	702343	1965
1966	0	0	0	0	0	0	0	0	0	0	0	0	1966
1967	3763	1446	101557	36465	45361	101769	112105	124880	120362	100114	619	748441	1967
1968	3450	1026	103470	36161	46039	103930	109682	106919	119608	99457	615	730357	1968
1969	3435	1398	115950	37982	47999	108536	117301	141056	127242	108936	683	810518	1969
1970	2904	1386	119531	38979	51321	115685	121332	147334	134352	114301	607	847732	1970
1971	2425	995	120619	39947	53333	119451	122858	151292	138665	117540	575	867700	1971
1972	2321	840	121490	40726	53849	119460	122963	150306	143194	120150	657	875965	1972
1973	2268	1136	121290	42145	54020	121313	124627	149392	140914	118843	668	876616	1973
1974	2266	1207	119023	38494	52754	120025	130822	156546	148836	121774	811	892558	1974
1975	2339	653	127442	39210	56388	124038	134064	162429	149324	120258	793	916938	1975

Sources: See notes following Table B-23.

Table B-6. Partially Irrigated Paddy, 1955-75

(hectares)

Year	Seoul	Pusan	Kyonggi	Kangwon	North Chung-chong	South Chung-chong	North Jolla	South Jolla	North Kyong-sang	South Kyong-sang	Jeju	South Korea	Year
1955	589	0	34808	10810	20204	38134	33750	52252	50481	40026	172	281226	1955
1956	589	0	35334	11233	18914	40941	33750	51997	48218	41421	156	282553	1956
1957	648	0	35302	10778	21325	41181	33750	43265	51627	43191	156	281223	1957
1958	624	0	36248	11227	20802	41437	31680	42415	50139	43056	170	277798	1958
1959	603	0	38960	10695	20505	42851	33750	42415	51665	40080	223	281747	1959
1960	615	0	38150	10880	19710	45067	32179	43674	51482	41133	190	283080	1960
1961	729	0	38302	10810	19014	44688	33049	45045	51351	40539	256	283783	1961
1962	588	0	38797	10781	19505	43656	32432	43356	48885	40748	156	278904	1962
1963	1915	756	39517	11681	19400	43971	29035	44714	50469	40131	183	281772	1963
1964	2009	1616	41499	10774	18974	44311	29260	45178	51881	40449	230	286181	1964
1965	2180	1542	42380	11186	20002	44375	29416	50889	52850	43618	262	298700	1965
1966	0	0	0	0	0	0	0	0	0	0	0	0	1966
1967	1921	1024	41781	10791	16966	39871	29080	47777	47663	42095	185	279154	1967
1968	1579	945	40941	10526	16538	39211	26009	34002	45227	37227	146	252351	1968
1969	1673	818	33129	10330	15877	38198	25741	42679	42046	37290	139	247920	1969
1970	1295	565	30123	8944	14007	33107	23578	38966	38559	33925	156	223225	1970
1971	977	570	29393	8683	12698	31277	22762	36303	36335	30954	148	210100	1971
1972	838	627	30212	8521	12529	35089	22952	34935	36817	29835	160	212515	1972
1973	1159	288	29970	7676	13971	34132	21334	35862	35822	29846	159	210219	1973
1974	1346	510	47183	14792	18953	51056	26534	46599	42644	38778	213	288608	1974
1975	1056	899	41816	15889	18713	50015	26030	41709	42448	41882	203	280660	1975

Sources: See notes following Table B-23.

(hectares)

Table B-7. Rain-irrigated Paddy, 1955-75

Year	Seoul	Pusan	Kyonggi	Kangwon	North Chung-chong	South Chung-chong	North Jolla	South Jolla	North Kyong-sang	South Kyong-sang	Jeju	South Korea	Year
1955	351	0	50219	8678	15312	39609	34976	49970	48079	29597	234	277025	1955
1956	351	0	51565	7659	15062	39609	35593	51205	47786	29571	234	278635	1956
1957	527	0	48948	7395	14911	40561	35593	39057	41189	32357	230	260768	1957
1958	527	0	47770	7358	13123	38176	37071	39853	41218	32588	35	257719	1958
1959	349	0	46995	7465	13078	37888	35646	39853	39946	29736	150	251106	1959
1960	344	0	46785	7287	12940	32598	32799	37978	39594	32791	135	243251	1960
1961	300	0	44969	6695	12840	31907	28379	33712	38440	34811	131	232184	1961
1962	205	0	38771	5818	11628	31047	29547	33548	34734	34296	157	219751	1962
1963	984	1025	41829	5270	12018	28331	23889	30312	36462	32815	165	213100	1963
1964	1220	1453	40297	5725	12027	26335	23235	30616	36412	27633	121	205074	1964
1965	1771	1313	36385	6077	12513	25584	22402	30929	33921	26874	81	197850	1965
1966	0	0	0	0	0	0	0	0	0	0	0	0	1966
1967	1492	1129	29027	5202	9986	23296	20938	31546	29920	23991	146	176673	1967
1968	1466	1087	26212	4986	9683	22125	17421	12582	30018	18489	169	144238	1968
1969	983	504	21720	4820	8694	19231	17581	22237	26053	17603	158	139584	1969
1970	736	277	18772	4337	6779	16126	14213	19100	19492	12563	149	112544	1970
1971	627	269	17783	3722	5740	13172	12388	17010	17805	11513	121	100150	1971
1972	779	299	16655	3331	5261	10975	11831	15516	13809	10703	131	89290	1972
1973	432	16	16228	3105	4997	10448	11392	14539	11516	10093	97	82863	1973
1974	127	151	13642	3826	7265	8061	10580	11596	17559	14914	42	87763	1974
1975	173	119	13418	2681	4312	6182	10037	10857	17746	13393	65	78983	1975

Sources: See notes following Table B-23.

Table B-8.   *Cultivated Upland (Dry Fields), 1952-75*

(hectares)

Year	Seoul	Pusan	Kyonggi	Kangwon	North Chung-chong	South Chung-chong	North Jolla	South Jolla	North Kyong-sang	South Kyong-sang	Jeju	South Korea	Year
1952	4080	0	102337	78364	72471	77453	61641	122079	153105	81629	35137	788296	1952
1953	4055	0	101582	77025	70855	78127	60934	122048	152694	83325	35821	786466	1953
1954	3996	0	103241	72820	73058	81738	61444	121163	152248	82445	36001	789154	1954
1955	3980	0	106921	82645	74462	82057	62126	122094	152292	82808	37378	806763	1955
1956	3887	0	106374	82743	76433	83909	63398	112796	152088	82859	37978	802465	1956
1957	3860	0	106079	82706	76547	84537	63262	113954	153379	82830	38138	805292	1957
1958	3815	0	107357	82905	76502	84307	63777	115908	154520	82717	40234	812042	1958
1959	3439	0	107283	82790	76668	85779	63067	116908	153865	82756	40291	812846	1959
1960	3414	0	107803	83475	76812	87253	63404	117763	153813	82586	41769	818092	1960
1961	2887	0	108324	82345	77215	88118	64190	120383	153049	82515	42195	821221	1961
1962	7713	1520	106198	80023	79630	97429	60866	124907	158134	80895	42877	840192	1962
1963	7624	1547	106122	85103	78998	93441	68672	129006	156381	81156	43485	851535	1963
1964	7974	1802	112744	91489	83173	100779	77892	141478	162589	85342	44605	909867	1964
1965	7685	1774	117087	100898	88707	108899	82967	155027	170673	89682	46782	970183	1965
1966	7142	2080	122026	105596	96543	112464	84299	159645	175845	92647	47672	1005959	1966
1967	6638	2014	122577	109206	99689	114127	86825	161318	177789	92476	48728	1021387	1967
1968	6134	1927	122261	111284	101127	116430	87987	161811	178799	92381	49302	1029443	1968
1969	5430	1866	121617	111706	101184	116942	88675	161500	178797	91830	48588	1028135	1969
1970	0	0	0	0	0	0	0	0	0	0	0	0	1970
1971	3528	1255	120247	108472	99841	116319	84403	155782	177028	91184	48398	1006457	1971
1972	3489	1214	120077	105273	98448	113230	83226	146890	173997	89575	47419	982838	1972
1973	4075	1120	119515	99444	98710	113762	83242	147080	174203	89824	47635	978610	1973
1974	3607	1098	119700	98481	97014	112706	81182	146504	173355	88032	47798	969477	1974
1975	3351	1046	117588	97142	97018	111695	80187	146744	172150	87728	48436	963087	1975

Sources: See Notes following Table B-23.

Table B-9.   Share of Double-cropped Paddy, 1952-75

(per mil)

Year	Seoul	Pusan	Kyonggi	Kangwon	North Chung-chong	South Chung-chong	North Jolla	South Jolla	North Kyong-sang	South Kyong-sang	Jeju	South Korea	Year
1952	4	0	45	16	249	150	323	354	441	460	272	290	1952
1953	4	0	39	18	255	161	334	339	440	473	266	302	1953
1954	0	0	34	32	246	170	343	361	446	507	404	303	1954
1955	14	0	34	25	311	322	488	500	452	538	258	374	1955
1956	0	0	35	29	257	200	449	355	467	565	394	330	1956
1957	0	0	33	21	259	199	386	361	471	579	415	324	1957
1958	0	0	36	23	258	198	411	371	484	591	367	333	1958
1959	0	0	33	23	243	199	377	373	499	596	434	331	1959
1960	0	0	31	24	237	197	364	380	509	600	322	331	1960
1961	0	0	29	26	246	213	504	443	510	618	398	367	1961
1962	7	454	29	17	252	241	496	491	528	633	362	382	1962
1963	5	437	30	29	261	246	509	509	568	635	359	394	1963
1964	7	437	32	36	296	258	499	546	603	681	361	414	1964
1965	7	497	95	42	350	326	503	588	623	716	407	451	1965
1966	8	466	48	53	365	341	513	608	642	730	406	458	1966
1967	10	459	70	51	365	344	546	636	653	738	401	474	1967
1968	9	460	111	51	364	353	569	672	664	747	392	493	1968
1969	6	461	124	48	368	350	572	690	670	752	356	499	1969
1970	8	462	137	46	352	332	552	687	668	751	382	495	1970
1971	9	482	156	44	348	318	540	701	672	753	350	497	1971
1972	8	444	157	41	329	306	528	687	667	753	428	489	1972
1973	0	440	125	40	312	303	511	684	654	749	424	477	1973
1974	409	551	158	50	487	527	728	784	621	793	521	573	1974
1975	194	506	257	68	455	565	764	798	616	819	528	599	1975

Sources: See notes following Table B-23.

Table B-10. Farm Population Density, 1952-75

(persons/100 hectares)

Year	Seoul	Pusan	Kyonggi	Kangwon	North Chung-chong	South Chung-chong	North Jolla	South Jolla	North Kyong-sang	South Kyong-sang	Jeju	South Korea	Year
1952	916	0	485	549	642	696	609	682	684	843	574	657	1952
1953	971	0	505	556	663	695	653	715	678	845	615	673	1953
1954	891	0	502	589	659	693	650	701	672	855	573	670	1954
1955	886	0	489	562	644	684	629	722	677	835	556	662	1955
1956	843	0	509	565	641	686	639	753	675	833	561	670	1956
1957	797	0	520	571	642	690	636	768	678	832	570	675	1957
1958	769	0	521	578	644	693	640	753	705	830	555	678	1958
1959	748	0	536	599	668	705	669	776	722	835	545	695	1959
1960	750	0	553	620	684	729	680	800	738	858	540	714	1960
1961	837		551	626	692	722	670	788	731	852	542	708	1961
1962	743	1077	557	619	711	748	688	808	750	869	561	727	1962
1963	715	1078	561	656	717	739	722	801	765	869	564	734	1963
1964	693	949	548	645	693	725	690	778	754	858	553	716	1964
1965	711	1015	541	590	673	706	697	757	731	855	545	700	1965
1966	711	1010	525	577	638	697	701	757	714	832	523	688	1966
1967	724	1044	529	576	649	701	715	772	718	837	541	695	1967
1968	691	1031	521	556	633	691	704	774	705	831	537	686	1968
1969	664	989	510	539	617	676	688	774	687	824	547	674	1969
1970	702	1053	505	541	605	662	686	766	679	805	536	665	1970
1971	740	1122	493	534	580	637	673	745	661	783	520	647	1971
1972	738	1108	495	536	588	643	678	769	662	783	542	654	1972
1973	691	1114	496	558	583	637	677	768	660	775	545	653	1973
1974	481	1071	475	560	523	593	634	692	593	699	504	601	1974
1975	456	731	447	469	502	542	581	596	567	630	466	547	1975

Sources: See notes following Table B-23.

Table B-11. Sale of Fertilizer Elements, 1952-75

(M/T)

Year	Seoul	Pusan	Kyonggi	Kangwon	North Chung-chong	South Chung-chong	North Jolla	South Jolla	North Kyong-sang	South Kyong-sang	Jeju	South Korea	Year
1952	230	0	12913	4264	6865	10644	12534	16431	19077	14930	1745	99633	1952
1953	266	0	13328	4423	7298	11120	12433	16835	19440	15218	1853	102225	1953
1954	299	0	14747	4871	8213	12476	13272	18778	21046	16887	1997	112609	1954
1955	301	0	15790	5506	8916	14264	14896	21081	23110	18233	2156	124284	1955
1956	324	0	16593	5744	9539	15745	16005	22893	24867	19453	2251	133454	1956
1957	386	0	17388	6104	10097	16963	16903	24360	26594	20679	2317	141834	1957
1958	395	0	17434	6237	10258	17410	17085	24984	27132	20811	2377	144170	1958
1959	368	0	17183	5983	10013	17409	16918	24683	26518	20495	2227	141849	1959
1960	336	0	16992	6054	9934	17422	16866	24560	26101	20212	2188	140721	1960
1961	422	0	21204	7880	12463	21903	21817	31113	33262	25340	2805	178266	1961
1962	433	0	21706	7871	14597	23862	24421	36196	37747	27790	3531	198213	1962
1963	3267	929	38098	18126	24456	42240	41768	62595	62842	45743	6905	346969	1963
1964	2858	952	38179	20029	25177	44854	43906	68848	67583	52448	6928	371762	1964
1965	2851	1119	38800	20857	26500	46771	48210	75199	63375	61223	8193	393098	1965
1966	2534	1034	43775	23115	27831	52341	51317	82913	72974	57250	8187	423271	1966
1967	3179	1176	50971	29054	32450	56899	57126	98782	83053	63802	9999	486491	1967
1968	3628	1179	51332	28726	31169	57597	55679	94085	81140	63146	10779	478460	1968
1969	3723	1269	55958	31825	35476	67317	62856	105338	89385	69584	11958	534689	1969
1970	3774	1321	60745	32703	35829	64631	65566	110759	99169	76152	12252	562901	1970
1971	3611	2593	62103	35293	47558	72244	79440	108236	108019	71410	14630	605137	1971
1972	4029	2953	67345	40106	40342	69284	74724	130987	115001	86095	16836	647702	1972
1973	4247	2596	84857	48968	54736	100342	103478	140201	137353	97331	19099	793208	1973
1974	3525	857	94989	51334	56346	102480	109053	150852	138858	106951	21414	836659	1974
1975	4109	1097	97775	49581	61816	113374	109858	158014	151169	116541	22873	886208	1975

Sources: See notes following Table B-23.

Table B-12. Sale of Pesticides, 1952-75

(M/T)

Year	Seoul	Pusan	Kyonggi	Kangwon	North Chung-chong	South Chung-chong	North Jolla	South Jolla	North Kyong-sang	South Kyong-sang	Jeju	South Korea	Year
1952	0	0	34583	4026	31855	36854	24415	29711	419740	99466	0	680650	1952
1953	35539	0	90968	20627	71269	73191	77958	108622	214085	82009	1146	775414	1953
1954	104943	0	187768	50649	63724	107814	128530	169858	690378	174125	261	1678050	1954
1955	286444	0	182723	27583	146442	187963	371678	745089	1727340	348105	617	4023984	1955
1956	236485	0	593662	208445	291323	253851	680257	1094399	1107319	354879	1773	4822393	1956
1957	335249	0	703413	157299	264756	858713	909730	1033422	1571407	920065	3562	6757616	1957
1958	531762	0	586571	97970	131405	544750	544969	683008	1304118	681108	1200	5106861	1958
1959	361614	0	375826	183212	136630	221791	449732	841859	1797223	1195001	7302	5570190	1959
1960	316740	0	292818	94633	43234	273939	360050	511581	2340723	1646911	80	5880709	1960
1961	279600	0	485394	45775	464398	368646	527025	448324	1701795	1250284	2437	5573678	1961
1962	884692	0	142263	151854	292292	385423	839276	670853	1185485	623518	9259	5184915	1962
1963	659534	320654	2341234	561625	1049048	2485551	2224316	2610713	3779133	2709266	31055	18772129	1963
1964	662826	507816	3539613	773834	1494139	3483192	3042591	2711570	4039293	2934532	166088	23355494	1964
1965	258710	60856	1226606	337581	856595	1682906	1524053	2302384	3150145	1301983	26882	12728701	1965
1966	1911003	365304	438143	169195	1095196	754872	2297665	2677686	1956491	879476	4444	12549475	1966
1967	1177317	240098	1193265	443279	653560	1324655	1323457	1332221	1678545	575734	46493	9988624	1967
1968	1028879	290437	850107	281485	937638	1666730	1459745	888378	2064858	485867	28850	9982972	1968
1969	1229441	248578	1840658	634296	1336952	3059710	2463864	1984071	3418599	1301717	13083	17530969	1969
1970	1709783	398180	3115256	649485	1209749	3119039	3117901	4001669	5186823	2504250	11526	25023661	1970
1971	3353819	880528	2636608	722636	1698451	2576259	4064401	4938049	6272452	2792391	24559	29960153	1971
1972	3119955	983288	2891193	1115233	2618653	2887365	3902717	5514967	7615539	2490801	58033	33197744	1972
1973	2599087	1262887	2423642	1027470	2994721	3911422	5033303	6709030	7142375	2813587	480202	36397726	1973
1974	4417756	1611400	5635111	1638045	4413915	8530308	7616290	10591154	11336099	6030431	781836	62602345	1974
1975	3311418	1583876	7550768	2811394	5331555	9423785	12370450	15408953	12638812	9466542	1071762	80969315	1975

Sources: See notes following Table B-23.

Table B-13.   Crop Coverage, 1941-75

(Number of Crops)

Year	Rice	Barleys	Misc. Grains	Pulses	Potatoes	Special Crops	Fruits	Vegetables	Livestock	All Crops	Year
1941	1	5	6	7	2	11	5	12	0	49	1941
1942	1	5	6	7	2	10	5	12	0	48	1942
1943	1	5	6	7	2	11	5	12	0	49	1943
1944	1	5	6	7	2	11	5	11	0	48	1944
1945	1	5	6	7	2	9	5	16	9	60	1945
1946	1	5	6	7	2	9	5	18	9	62	1946
1947	1	5	6	7	2	10	6	17	8	62	1947
1948	1	5	6	7	2	10	6	17	9	63	1948
1949	1	5	6	7	2	10	6	17	9	63	1949
1950	1	5	6	7	2	10	6	17	8	62	1950
1951	2	5	6	7	2	7	0	0	8	37	1951
1952	2	5	6	7	2	9	7	19	9	66	1952
1953	2	5	6	7	2	10	7	19	9	67	1953
1954	2	5	6	7	2	10	7	19	9	67	1954
1955	2	5	6	7	2	12	7	19	9	69	1955
1956	2	5	6	7	2	15	7	20	9	73	1956
1957	2	5	6	7	2	15	7	20	9	73	1957
1958	2	5	6	7	2	15	7	20	9	73	1958
1959	2	5	6	7	2	14	7	16	9	68	1959
1960	2	5	6	7	2	15	7	19	9	72	1960
1961	2	4	6	7	2	16	7	20	10	74	1961
1962	2	4	6	7	2	17	7	18	10	73	1962
1963	2	4	6	7	2	19	7	18	10	75	1963
1964	2	4	6	7	2	20	7	18	10	76	1964
1965	2	4	6	7	2	21	7	19	10	78	1965
1966	2	4	6	7	2	21	7	20	10	79	1966
1967	2	4	6	7	2	20	7	18	10	76	1967
1968	2	4	6	7	2	21	8	22	0	72	1968
1969	2	4	6	7	2	20	8	22	10	81	1969
1970	2	4	6	7	2	19	8	22	5	75	1970
1971	2	4	6	7	2	20	8	23	10	82	1971
1972	2	4	6	7	2	19	8	22	10	80	1972
1973	2	4	6	7	2	17	8	23	10	79	1973
1974	2	4	6	7	2	17	8	22	10	78	1974
1975	2	4	6	7	2	18	8	22	10	79	1975

Sources: See notes following Table B-23.

(M/T)

Table B-14. Total Rice Production, 1941-75

Year	Seoul	Pusan	Kyonggi	Kangwon	North Chung-chong	South Chung-chong	North Jolla	South Jolla	North Kyong-sang	South Kyong-sang	Jeju	South Korea	Year
1941	0	0	379653	162048	165463	384769	429533	445598	431424	409332	0	2807820	1941
1942	0	0	360607	186168	130614	324255	417878	222791	296943	178103	0	2117359	1942
1943	0	0	247442	179869	163786	384811	510149	453632	438713	369258	0	2747660	1943
1944	0	0	278788	209142	129297	296720	397833	416844	161914	198985	0	2089523	1944
1945	0	0	373819	84283	169346	345551	383124	485707	401541	316745	0	2560116	1945
1946	0	0	361765	83995	138893	280179	343878	436832	424614	327018	2866	2400040	1946
1947	1113	0	396203	87549	160149	348282	403125	477454	510913	375076	4336	2764200	1947
1948	1175	0	505387	95645	175537	362756	422112	496572	573410	442984	5165	3080743	1948
1949	1184	0	327961	81598	150628	286778	463121	578260	580957	469025	5555	2945067	1949
1950	1157	0	386436	81397	147218	298297	447193	550168	554144	441770	5780	2913560	1950
1951	988	0	293462	56968	123238	333052	373008	456796	264922	352943	3015	2258392	1951
1952	5593	0	250727	58850	102501	254115	299692	437890	209572	225790	2088	1846818	1952
1953	6148	0	387820	88294	162439	387267	373892	477040	526178	409386	2911	2821375	1953
1954	6549	0	398161	79261	153800	392240	436697	575180	573148	381984	3331	3000351	1954
1955	6655	0	400179	122272	149492	374843	441989	577229	570798	447238	3291	3093986	1955
1956	5465	0	365693	80211	147250	367620	381517	416495	473558	314237	2214	2554260	1956
1957	5995	0	412268	112948	180138	414608	443892	597760	516441	451248	2400	3137698	1957
1958	6340	0	450504	97909	187889	452101	474029	580375	565358	491086	3345	3308936	1958
1959	6231	0	466250	124669	200287	477270	489407	571049	552864	422004	2756	3312787	1959
1960	6667	0	478045	132648	208459	489696	481006	542319	440720	391909	3264	3174733	1960
1961	6587	0	507076	137331	226337	524882	537889	656558	631556	536935	4268	3769419	1961
1962	5609	0	429592	155031	168992	455677	426636	581007	500459	388690	5013	3116706	1962
1963	27270	11762	518174	133538	214093	541661	545698	681433	630460	504152	5719	3813960	1963
1964	27182	15024	551205	144449	236485	559401	559126	720615	643887	551376	4754	4013504	1964
2965	21172	12531	417702	116186	197854	506792	489321	604161	590510	538364	6529	3501122	1965
1966	19366	13547	516137	114960	224530	537303	560839	705862	653286	569036	4205	3919271	1966
1967	21570	10966	544219	130826	212858	549878	498172	469431	629982	531650	3542	3603094	1967
1968	18909	6829	486615	119802	192062	614209	454872	381878	494720	421395	4035	3195326	1968
1969	18988	8101	596226	150842	210288	597138	585175	719003	672057	528126	4491	4090435	1969
1970	14971	6410	607384	129421	220590	611574	530748	637029	663094	515328	2702	3939251	1970
1971	12577	5673	572342	152600	250898	558141	550920	720266	618377	552623	3210	3997627	1971
1972	6459	5443	540879	153778	224777	572907	567018	714778	625798	540667	4677	3957181	1972
1973	13352	3784	638896	186021	271345	607194	534401	741608	703157	506776	5084	4211618	1973
1974	10308	3531	662163	150313	299593	653771	627096	689294	736198	607534	5091	4444847	1974
1975	11982	4127	700516	207318	342554	678985	637071	767353	815692	497901	5588	4669087	1975

Sources: See notes following Table B-23.

Table B-15. Rice Yield Correction Factors, 1962-64

(per mil)

Year	Seoul	Pusan	Kyonggi	Kangwon	North Chung-chong	South Chung-chong	North Jolla	South Jolla	North Kyong-sang	South Kyong-sang	Jeju	South Korea	Year
1962	0	0	1198	1552	1405	1491	1321	1310	1490	1087	1344	1314	1962
1963	1312	1403	1237	1255	1429	1351	1377	1461	1430	1286	1397	1360	1963
1964	1375	1609	1269	1351	1409	1313	1279	1350	1427	1480	1408	1355	1964

Sources: See notes following Table B-23.

Table B-16. Farm Output (1970 Prices), 1941-75

(million 1970 won)

Year	Seoul	Pusan	Kyonggi	Kangwon	North Chung-Chon	South Chung-chong	North Jolla	South Jolla	North Kyong-sang	South Kyong-sang	Jeju	South Korea	Year
1941	0	0	45227	21727	21947	50555	44795	52697	54838	48124	0	339914	1941
1942	0	0	47731	23700	21056	41403	44193	34623	43199	29491	0	285400	1942
1943	0	0	36848	24814	23692	47394	51350	54511	56091	45939	0	340643	1943
1944	0	0	39802	28210	23965	41268	44654	54001	38740	36440	9	367083	1944
1945	1758	0	46026	13587	20720	36456	38685	56126	49735	42620	9	305726	1945
1946	742	0	48698	14524	21490	33293	37426	55481	56027	45530	2979	316195	1946
1947	966	0	52836	14984	20580	39292	42726	54498	61300	49089	4406	340681	1947
1948	923	0	64789	16762	24332	45253	44123	61749	75923	58493	5438	397789	1948
1949	2392	0	58046	17087	25190	38113	49486	71641	81342	62583	5137	411021	1949
1950	1595	0	53356	13662	21205	33353	44884	63352	69890	56443	4694	364437	1950
1951	115	0	29746	7771	16412	35793	35619	50973	39798	44169	3853	264255	1951
1952	2680	0	41433	11333	19174	37276	41774	58991	48770	40769	3634	305840	1952
1953	1894	0	50459	15536	25200	49046	50029	67386	74875	54552	5144	394125	1953
1954	3498	0	52850	14904	25917	52915	56903	78594	81016	58942	5613	431156	1954
1955	3121	0	51374	21962	24655	52814	52333	83814	82921	65755	5849	444603	1955
1956	2343	0	47131	18451	25099	50027	52652	68452	75626	53376	5091	398253	1956
1957	3044	0	52186	22921	27238	52460	57358	79936	78920	62910	5278	442255	1957
1958	3797	0	59849	21918	28921	57878	59376	83043	83038	69750	5398	472973	1958
1959	4202	0	62002	22863	30561	61075	58676	81281	87706	64733	6072	479176	1959
1960	4068	0	62197	22626	31744	62797	57122	77875	77333	61828	6944	464537	1960
1961	3679	0	67227	24904	37566	67496	66195	95408	96673	77599	7140	543891	1961
1962	3477	0	68259	28492	35690	66421	60214	92600	91978	71246	10312	529246	1962
1963	6099	554	66607	28929	34875	67653	63637	81189	89985	64561	6938	508048	1963
1964	6394	1571	76263	29178	43689	77487	73692	103215	111162	83756	9544	616582	1964
1965	6240	2197	61320	26109	41164	72691	71638	100671	112704	100297	11474	606657	1965
1966	5839	2345	75971	30143	50036	86745	86582	120471	127650	100929	12364	699257	1966
1967	6575	2521	78949	34796	52512	88632	74938	91542	123196	98416	10220	661879	1967
1968	5842	2097	77444	33744	54107	94069	79190	100430	121614	97601	12929	678884	1968
1969	5674	1909	90080	34214	51806	90552	88309	124695	132777	114222	11677	745957	1969
1970	5080	1946	92849	32543	53497	89725	82585	116570	130806	106353	11014	722950	1970
1971	4351	1923	96469	37210	56618	86926	85110	127748	134152	110017	11219	751904	1971
1972	4264	2078	88892	37201	56858	90079	87892	131958	139982	107624	13244	760192	1972
1973	5177	2194	101424	40841	63412	95021	82911	132537	144259	101105	17617	786211	1973
1974	4462	1900	109094	40270	66800	100575	88305	128285	149282	109316	20264	818692	1974
1975	4236	1863	113478	43831	68785	102920	89640	135066	157276	106509	36805	860414	1975

Sources: See notes following Table B-23.

Table B-17. Farm Output (Current prices), 1959-75

(million 1970 won)

Year	Seoul	Pusan	Kyonggi	Kangwon	North Chung-chong	South Chung-chong	North Jolla	South Jolla	North Kyong-sang	South Kyong-sang	Jeju	South Korea	Year
1959	4795	0	47541	19160	23644	46141	43913	61819	68394	49157	5241	369809	1959
1960	4361	0	52313	19785	26969	52703	47810	65220	67305	54374	6801	397645	1960
1961	3001	0	58956	21642	33783	60974	59779	84264	87759	69888	6284	486335	1961
1962	2596	453	58938	25289	33051	59869	54093	84449	87396	65353	10157	481649	1962
1963	6634	1739	79638	28773	42710	82176	77165	96505	108853	76174	8139	608513	1963
1964	6756	2764	93262	34631	56777	96817	92612	130932	146653	109901	12747	783858	1964
1965	5915	2476	62621	27213	43626	75880	73440	104274	120398	107637	12760	636246	1965
1966	5364	2464	74814	30399	51165	86805	84686	116515	127650	99106	12836	691811	1966
1967	6173	2156	74905	35566	49357	85428	70984	88662	120068	96373	10880	640558	1967
1968	5000	1856	67637	30834	44101	84079	70908	92709	108099	90038	12968	608235	1968
1969	4950	1867	84398	31826	46194	85878	84622	120586	127312	107779	11433	706849	1969
1970	5080	1923	92849	32543	53497	89725	82585	116570	130806	106353	11014	722950	1970
1971	4227	2095	96825	37820	57616	90512	90491	138070	141779	116566	12024	788030	1971
1972	3641	2022	93446	40106	61662	100575	100942	155161	157565	119871	14217	849214	1972
1973	4859	1821	107046	45897	66845	104511	94209	153697	161490	120020	17666	870067	1973
1974	4286	1979	118487	44491	71259	113870	103246	149688	169835	124481	18033	919661	1974
1975	4284	1929	127215	47929	71110	118566	105955	162598	177476	123054	27991	968113	1975

Sources: See notes following Table B-23.

Table B-18. Per-Household Farm Output, 1941–75

(thousand 1970 won)

Year	Seoul	Pusan	Kyonggi	Kangwon	North Chung-chong	South Chung-chong	North Jolla	South Jolla	North Kyong-sang	South Kyong-sang	Jeju	South Korea	Year
1941	0	0	251	247	160	219	184	128	161	163	0	159	1941
1942	0	0	264	275	155	178	180	83	129	100	0	134	1942
1943	0	0	193	237	157	192	190	123	157	144	0	147	1943
1944	0	0	201	288	176	172	174	126	110	118	0	137	1944
1945	0	0	164	150	169	157	160	135	145	142	0	151	1945
1946	476	0	172	137	164	142	145	139	159	144	128	151	1946
1947	309	0	186	123	148	166	156	141	171	149	95	156	1947
1948	117	0	211	132	163	175	159	153	195	163	115	171	1948
1949	189	0	175	129	159	135	175	170	193	160	107	166	1949
1950	126	0	224	116	142	131	178	153	179	150	103	161	1950
1951	10	0	126	67	109	131	142	125	100	119	84	117	1951
1952	255	0	179	100	127	136	168	147	121	113	79	136	1952
1953	176	0	219	138	165	178	194	170	183	150	111	175	1953
1954	327	0	228	135	170	192	222	199	203	162	119	192	1954
1955	307	0	220	170	162	194	214	212	216	183	130	200	1955
1956	240	0	199	141	168	187	213	176	199	153	112	181	1956
1957	339	0	217	176	181	196	231	203	208	181	112	200	1957
1958	458	0	251	167	194	221	240	210	210	200	114	213	1958
1959	532	0	253	168	201	226	233	203	216	182	127	211	1959
1960	549	0	244	160	198	221	221	187	185	170	139	197	1960
1961	538	0	264	174	233	241	261	230	235	218	140	233	1961
1962	170	55	266	202	209	211	233	210	207	196	187	214	1962
1963	320	156	265	166	208	231	242	190	209	183	123	210	1963
1964	339	222	299	186	257	261	276	238	258	233	168	251	1964
1965	336	229	236	165	237	238	259	228	255	273	196	241	1965
1966	332	250	291	189	283	279	308	266	284	273	218	275	1966
1967	397	218	299	215	290	278	253	198	273	265	174	255	1967
1968	406	211	297	212	300	294	268	215	271	264	221	263	1968
1969	447	235	352	221	292	286	302	267	300	313	201	292	1969
1970	445	253	364	214	307	287	288	252	299	295	190	287	1970
1971	462	306	381	247	333	281	301	278	311	309	195	303	1971
1972	464	347	354	256	336	295	315	290	328	306	231	310	1972
1973	541	343	405	281	374	312	297	291	339	290	304	320	1973
1974	787	363	428	266	417	335	322	294	368	326	367	343	1974
1975	657	377	432	317	423	348	330	312	377	318	660	361	1975

Sources: See notes following Table B-23.

Table B-19. Per-Household Rice Output, 1941-75

(1970 won)

Year	Seoul	Pusan	Kyonggi	Kangwon	North Chung-chong	South Chung-chong	North Jolla	South Jolla	North Kyong-sang	South Kyong-sang	Jeju	South Korea	Year
1941	0	0	161600	141281	92762	127677	134990	82938	97363	106632	0	100937	1941
1942	0	0	153117	165779	73964	106684	130324	41183	67921	46499	0	76276	1942
1943	0	0	99206	131432	83508	119491	145077	78625	94244	89098	0	90894	1943
1944	0	0	107852	163676	72741	95035	119230	74663	35446	49357	0	71388	1944
1945	0	0	101992	70984	106264	114270	121357	90005	89618	80831	0	96895	1945
1946	0	0	98297	60612	81504	91620	102350	83812	92661	79638	9454	87551	1946
1947	27286	0	107213	55278	88255	112634	113153	94965	109068	87350	7151	97325	1947
1948	11433	0	125910	57863	90384	107309	116551	94333	112686	94452	8393	101434	1948
1949	7186	0	75701	47382	72781	78167	125823	105131	105766	92216	8896	91060	1949
1950	7022	0	124139	53160	75716	84676	136263	101742	109034	89856	9759	98577	1950
1951	6547	0	95695	37990	63083	93833	144434	85863	51225	73293	5083	76868	1951
1952	40809	0	83014	40091	52220	71060	92574	83684	39834	47892	3514	63239	1952
1953	43851	0	129172	60350	81826	107873	111111	92070	98720	86447	4830	95980	1953
1954	46859	0	131432	55056	77515	109425	130898	111949	110248	80440	5442	102713	1954
1955	50097	0	131166	72814	75424	105851	138875	112045	113934	95705	5625	106731	1955
1956	42983	0	118596	47172	75551	105489	118091	81974	95627	69075	3739	88844	1956
1957	51109	0	131408	66707	92024	118840	137328	116133	104164	99331	3923	108623	1957
1958	58554	0	144729	57237	96690	132488	146613	127771	109801	108009	5407	114176	1958
1959	60406	0	146019	70182	101138	135553	149172	109487	104429	90968	4439	111826	1959
1960	68866	0	143931	71817	99682	132275	142400	99927	80933	82660	5021	103410	1960
1961	73693	0	152330	73784	107450	143807	162712	121369	117869	115850	6433	123963	1961
1962	21040	0	128173	84407	75815	110932	126510	101157	86313	82019	6953	96504	1962
1963	109639	89832	157842	68160	97740	141976	158865	122396	112509	109359	7757	120769	1963
1964	110516	116396	165387	70586	106475	144319	160471	127407	114633	117725	6404	125287	1964
1965	87292	93648	123290	56485	87181	127144	135663	104678	102359	112480	8547	106825	1965
1966	84246	103084	151538	55143	97463	132257	152869	119606	111464	118085	5676	118012	1966
1967	99643	87547	157682	61992	89999	132312	128764	77964	106929	109631	4634	106538	1967
1968	100580	57884	142803	57832	81622	147182	118043	62536	84445	87438	5297	94786	1968
1969	114650	75008	178248	74616	90935	144483	133535	117999	116443	110776	5913	122877	1969
1970	100379	64547	182232	65127	97080	149884	141601	105545	116142	109526	3580	119828	1970
1971	102320	63897	173155	77782	113199	138427	149387	120150	109733	118879	4287	123221	1971
1972	53788	65969	165194	81198	101846	143831	155767	120173	112308	117925	6247	123451	1972
1973	106818	52235	195415	97924	122710	152554	146437	124643	126411	111199	6715	131473	1973
1974	139156	48229	198955	76119	143374	166826	175025	121019	139036	138777	7058	142779	1974
1975	142181	63979	204152	115015	161337	175876	179456	135980	149808	113942	7668	150117	1975

Sources: See notes following Table B-23.

Table B-20. Per-Household Barley Output, 1941-75

(1970 won)

Year	Seoul	Pusan	Kyonggi	Kangwon	North Chung- chong	South Chung- chong	North Jolla	South Jolla	North Kyong- sang	South Kyong- sang	Jeju	South Korea	Year
1941	0	0	9530	8122	19523	18554	18428	17389	25055	21988	0	16893	1941
1942	0	0	21738	16393	35565	25177	21516	21902	34323	32560	0	24178	1942
1943	0	0	16297	11709	25488	24337	17628	15453	24068	24981	0	18252	1943
1944	0	0	26583	17883	49321	37095	25197	21293	48428	42288	0	30396	1944
1945	0	0	4264	3936	3719	6182	4129	10563	17731	25823	0	11188	1945
1946	0	0	11683	5564	25072	16622	14126	19952	23244	24796	0	18387	1946
1947	1945	0	12076	6105	17003	15077	12436	15215	19455	21740	16373	15741	1947
1948	147	0	9943	5450	20516	12211	12146	14114	19374	20598	47221	15560	1948
1949	138	0	16221	7189	24962	18713	13039	16593	25412	23508	27596	19128	1949
1950	135	0	20164	8240	27210	20300	16210	17673	26667	28236	23985	21557	1950
1951	62	0	10873	3245	14171	10894	8574	10170	16000	18401	21642	12621	1951
1952	2330	0	12653	5500	18677	14912	16310	17099	24435	21176	22752	17832	1952
1953	6412	0	15287	8891	25093	16342	20583	23136	23746	20201	28741	20308	1953
1954	8670	0	23117	10567	29236	23859	21559	27130	33999	27244	38735	26399	1954
1955	5605	0	18746	8298	16873	18934	16934	20959	31513	28564	36528	22318	1955
1956	4070	0	19766	8467	26538	21818	19035	25442	30806	24769	38400	23734	1956
1957	2651	0	14193	9519	21872	16145	16776	23678	28153	24116	30420	20947	1957
1958	4929	0	22149	10217	25757	23300	20204	27735	28381	31588	26532	25174	1958
1959	5432	0	23091	10069	31214	25371	23314	29689	36915	32373	37499	28475	1959
1960	5024	0	22791	9853	29461	24048	22762	26603	36994	32827	35549	27573	1960
1961	3317	0	23809	10599	33935	26540	25628	32287	38943	32990	36115	30010	1961
1962	367	0	26281	9572	26923	24283	25813	32908	39274	41436	48441	30981	1962
1963	5389	2752	10299	6692	12953	10362	9032	7270	12838	8145	9984	9654	1963
1964	9934	34351	28718	10029	33015	29491	29104	31092	50373	47715	25098	34746	1964
1965	3717	35821	15678	7629	28927	24490	34746	39341	51556	72847	46987	38815	1965
1966	3140	34259	24222	13763	44833	42298	33732	45737	53562	55310	38900	41990	1966
1967	1707	23106	20493	14262	40426	36700	27120	45183	50350	55011	42893	39190	1967
1968	1280	21248	17382	10019	32079	33965	42873	54693	46127	60746	46360	41454	1968
1969	1289	16485	23041	12827	37529	33767	41172	50272	54425	59435	31077	42526	1969
1970	427	12115	15560	11116	35792	29449	41351	53337	52442	55282	33752	40719	1970
1971	144	10521	12568	10666	31219	25545	36936	54471	47247	55330	33709	38439	1971
1972	43	8983	14494	10554	30510	27914	36243	63098	47811	53292	18438	39885	1972
1973	58	7664	14806	11210	28474	28166	29115	56393	40177	44729	26401	35413	1973
1974	187	6939	12844	10781	25957	26038	28312	53743	43790	47642	28355	34989	1974
1975	313	3090	12363	9170	22406	27131	29830	50673	44906	49452	28413	34873	1975

Sources: See notes following Table B-23.

Table B-21. Per-Household Vegetable Output, 1941-75

(1970 won)

Year	Seoul	Pusan	Kyonggi	Kangwon	North Chung-chong	South Chung-chong	North Jolla	South Jolla	North Kyong-sang	South Kyong-sang	Jeju	South Korea	Year
1941	0	0	49913	34575	17069	48363	14497	7411	10741	11860	0	8412	1941
1942	0	0	55303	34054	15436	24804	14198	6262	7470	8053	0	14915	1942
1943	0	0	38915	28168	12382	27855	10408	7885	9553	6129	0	13284	1943
1944	0	0	32869	41426	14582	20734	8417	7975	5630	7466	0	12196	1944
1945	0	0	25130	19316	16879	13859	8028	6333	5158	9174	0	12304	1945
1946	412512	0	34824	20944	25398	10888	9173	7244	6473	8208	7282	13729	1946
1947	240308	0	39421	18409	14084	15003	10774	8410	6281	11621	10085	14882	1947
1948	91771	0	43305	19261	15436	29348	8285	12988	24300	16997	9937	21472	1948
1949	156730	0	48223	18328	16130	10408	10877	13659	22054	15253	12169	20387	1949
1950	109711	0	46977	14424	12019	7585	8539	9726	16685	11044	12195	15679	1950
1951	0	0	0	0	0	0	0	0	0	0	0	0	1951
1952	189426	0	53659	22200	21303	18631	35602	17749	16211	15390	6214	23926	1952
1953	102283	0	40092	20052	21137	24094	31579	24315	23785	13737	9878	24580	1953
1954	218389	0	30703	22437	14982	18323	35001	23170	15771	21915	10412	23262	1954
1955	186396	0	27944	31390	14408	25071	12942	25135	18416	21695	10720	22445	1955
1956	141931	0	19249	28466	14035	19821	31228	19872	19692	20090	9241	21512	1956
1957	210073	0	26545	34801	18397	18835	36436	19544	22813	20387	18185	24364	1957
1958	274003	0	36075	32219	21847	22308	31303	21898	17881	22328	15759	25289	1958
1959	322191	0	32713	21795	17317	21251	23505	15582	15776	19681	19573	21296	1959
1960	350225	0	28987	20570	19176	24046	16933	15630	14194	16726	22143	19880	1960
1961	339171	0	35777	23224	24359	24247	28323	28002	17341	24520	10371	26013	1961
1962	73963	0	45934	31127	34659	23953	25588	15309	18469	21954	10581	24754	1962
1963	124209	9791	40738	25916	30848	27578	25431	10723	21227	17834	9832	23588	1963
1964	148646	16410	48617	36962	36411	31869	29888	21930	26034	20615	12640	30001	1964
1965	156230	30479	39882	29796	37100	27876	33976	24918	27469	27799	15788	30623	1965
1966	147125	34410	52524	44470	47998	37931	44504	29163	36286	30303	19548	38640	1966
1967	175226	28794	56912	48111	66854	42187	33729	23801	34565	36940	15225	39656	1967
1968	152344	37560	66152	48987	91150	43812	38314	28930	50870	43242	20957	47580	1968
1969	149304	28054	71750	39940	68940	39702	36706	27760	35437	63318	18784	45112	1969
1970	125541	26757	78484	38173	76911	36793	31532	24388	38257	56818	14455	43983	1970
1971	98707	32320	107841	54269	88783	50860	42900	30460	50497	56806	20185	54999	1971
1972	131009	64370	81223	56695	90253	51143	44807	30703	52384	55937	23862	53322	1972
1973	147968	27066	79029	52919	104657	52147	43249	32068	51891	53200	21975	53527	1973
1974	260659	31107	86717	56647	113006	54968	43597	35209	55851	55737	23115	57395	1974
1975	257339	60043	85451	64547	104755	57525	45635	37718	58420	71168	25600	61158	1975

Sources: See notes following Table B-23.

Table B-22. Per-Household Farm Output, 1959-75

(1000 deflated won)

Year	Seoul	Pusan	Kyonggi	Kangwon	North Chung-chong	South Chung-chong	North Jolla	South Jolla	North Kyong-sang	South Kyong-sang	Jeju	South Korea	Year
1959	607	0	194	141	156	171	174	154	168	138	110	163	1959
1960	588	0	205	140	168	186	185	157	161	149	136	169	1960
1961	439	0	231	152	209	218	236	203	214	197	123	209	1961
1962	127	45	229	180	193	190	209	192	197	180	184	194	1962
1963	348	173	317	192	254	281	293	226	253	216	144	251	1963
1964	359	280	365	221	334	326	347	302	341	306	224	319	1964
1965	318	241	241	172	251	248	266	236	272	294	218	253	1965
1966	305	245	287	190	290	279	301	258	284	268	226	272	1966
1967	372	225	283	220	272	268	239	192	266	259	186	247	1967
1968	347	205	259	194	245	263	240	198	241	244	222	235	1968
1969	390	226	329	205	261	271	290	258	288	295	196	277	1969
1970	445	253	364	214	307	287	288	252	299	295	190	287	1970
1971	449	308	382	252	339	293	320	301	328	327	209	317	1971
1972	396	320	373	276	365	330	362	341	369	341	248	346	1972
1973	508	328	428	315	395	343	337	337	379	321	305	355	1973
1974	756	353	465	294	445	379	376	343	419	371	326	386	1974
1975	664	390	484	347	437	401	390	376	426	368	502	406	1975

Sources: See notes following Table B-23.

Table B-23.   Per Capita Farm Output, 1941-75

(thousand 1970 won)

Year	Seoul	Pusan	Kyonggi	Kangwon	North Chung-chong	South Chung-chong	North Jolla	South Jolla	North Kyong-sang	South Kyong-sang	Jeju	South Korea	Year
1941	0	0	42	41	27	37	32	23	27	28	0	27	1941
1942	0	0	44	46	26	30	32	15	21	17	0	23	1942
1943	0	0	32	39	26	32	34	22	26	25	0	25	1943
1944	0	0	34	43	30	29	31	28	17	19	0	23	1944
1945	0	0	28	18	29	26	29	40	20	21	28	26	1945
1946	0	0	27	18	27	23	26	30	23	20	21	25	1946
1947	0	0	27	17	24	27	28	24	25	20	25	24	1947
1948	0	0	33	21	27	28	27	26	31	25	23	28	1948
1949	26	0	29	24	27	22	30	28	32	27	21	28	1949
1950	33	0	40	20	24	21	31	29	30	25	17	28	1950
1951	22	0	22	12	18	21	25	23	17	20	17	20	1951
1952	1	0	31	18	21	22	29	26	20	19	22	23	1952
1953	46	0	37	24	27	29	32	28	31	25	26	29	1953
1954	30	0	38	23	28	31	37	34	34	27	27	32	1954
1955	61	0	37	30	26	31	34	35	35	31	23	33	1955
1956	55	0	33	25	28	29	37	28	32	25	23	29	1956
1957	44	0	35	30	28	30	37	32	33	29	23	32	1957
1958	61	0	40	29	30	33	38	34	33	33	23	34	1958
1959	82	0	40	29	30	34	36	32	34	30	26	33	1959
1960	101	0	39	27	31	33	34	29	29	28	29	31	1960
1961	99	0	42	30	36	36	40	36	37	35	30	37	1961
1962	91	9	43	36	32	32	37	34	34	32	41	35	1962
1963	28	26	42	28	32	34	37	29	33	29	27	33	1963
1964	52	36	47	31	39	38	43	37	40	37	37	39	1964
1965	55	38	37	28	36	35	40	35	40	43	43	38	1965
1966	54	41	47	32	45	42	48	41	46	44	48	44	1966
1967	63	35	48	36	45	43	40	30	44	43	37	41	1967
1968	64	34	48	35	47	46	43	33	44	43	47	42	1968
1969	72	39	58	37	46	45	49	42	49	51	42	47	1969
1970	71	41	60	36	49	46	47	40	49	49	41	47	1970
1971	74	50	64	42	55	46	50	46	52	52	43	51	1971
1972	75	57	59	43	54	48	52	47	55	51	50	51	1972
1973	91	57	68	46	61	51	49	47	57	49	66	53	1973
1974	126	63	76	46	72	58	55	51	65	59	82	60	1974
1975	134	93	84	60	77	64	61	62	72	64	159	70	1975

Sources: See notes following this Table.

Notes to Tables in Appendix B

Table B-1

The figures in this table were taken from the Ministry of Agriculture and Forestry, *Yearbook of Agriculture and Forestry Statistics,* Seoul, individual years, 1952-1976, and from the Bank of Chosun, Research Division, *Chosun Kyeongje Yeonbo, 1948* [Chosun Economic Yearbook, 1948], Seoul, 1948 and *Kyeongje Yeongam, 1949* [Economic Review, 1949], Seoul, 1949. Administrative data for 1946 and for 1970 were not reported and for purposes of analysis in the text, estimates for these years were computed as the averages from data for preceding and following years.

Table B-2

Data in this table are from the same sources as in Table B-1, except that estimates for 1943 were also obtained by averaging.

Table B-3

The M.A.F. administrative data report the numbers of full-time and part-time households, depending on whether all or only a fraction of household income is from farming. In this table, full-time households as a share of all households are given in per mil (645 per mil = 64.5 percent).

Table B-4

Beginning in 1968 the M.A.F. reported the number of farm households living solely from farm-wage income. This table presents their number as a fraction of all farm households.

Table B-5

Paddy included in this table is either part of land irrigated by irrigation associations or securely irrigated by private individuals.

Tables B-6 and B-7

The data from these two tables are generally considered to represent insecurely irrigated paddy, and are actually referred to as "unirrigated paddy" in the text of this study.

Table B-8

Upland includes all non-paddy cultivated land, hence orchards and mulberry groves are included in this category.

Table B-9

Double-cropped paddy is paddy on which a winter crop of barley or winter crops of vegetables under vinyl greenhouses are planted after the autumn rice harvest. This table gives such paddy as a share of total paddy.

Table B-10

Figures in this table are calculated by dividing farm population data in Table B-1 by the sum of farm land data in Tables B-5, B-6, B-7, and B-8.

Tables B-11 and B-12

These data report levels of government sale of chemicals to farmers. Such levels are close to total consumption levels because of the government's near monopoly on such sales. Note, however, that the method of reporting changed in 1963, making comparisons over time difficult. For a discussion of the difficulties this introduces in the case of fertilizer, see the treatment on pages 152-154.

Table B-13

This table gives the number of output crops in each category for which crop data were available. These crops were the basis for value output calculations reported in Table B-16 and subsequent tables.

Table B-14

The data are for polished rice. For a discussion of procedures used in the preparation of this table see the commentary at the beginning of this appendix.

Table B-15

These data represent the ratio of yields reported by the revised sampling method to yields reported by the method in use before 1964. (Note that 1312 per mil represents a ratio of 1.312 to 1.000) The data were obtained from unpublished records of yield reports, Ministry of Agriculture and Forestry, Seoul.

Table B-16

Each of these figures represents $V_j = \sum_i P_{i70} Q_{ij}$, where V_j is total value output in some province for year j, P_{i70} is the price of the ith product in 1970 and Q_{ij} is the quantity output of the ith product in the year j.

Table B-17

Each of these figures represents $V_j = \sum_i Q_{ij} P_{ij}/C_j$, where V_j and Q_{ij} are as above, P_{ij} is the price of the ith product in the jth year and C_j is the farm consumer price index with $1970 = 1.000$.

Tables B-18, B-19, B-20 and B-21

These data represent the division of household data from Table B-2 into 1970 constant-price value data for all crops as reported in Table B-16 and for value output for three of its component categories, rice, barleys and vegetables.

Table B-22

These data are the result of dividing the current-price output data in Table B-17 by the number of farm households for appropriate years from Table B-2.

Table B-23

These figures were obtained by dividing farm population figures from Table B-1 into constant-price value output data from Table B-16.

APPENDIX C

PRICE AND TERMS OF TRADE STATISTICS: 1959-1975

This appendix presents the prices used to combine individual crop quantity output data into total value output figures. Each price represents the average of monthly average prices as reported by the National Agricultural Cooperatives Association (N.A.C.A.). In addition, this appendix gives the N.A.C.A.'s index for farm consumer prices, which reflects not only changes in the farm cost of living, but also changes in the prices of major farm inputs. In Table C-4 the movement of each price is compared with the movement of the average farm cost of living, while in Tables C-1 and C-2 the change of all prices as measured by a Paasche price index is compared to the change in farm cost of living, first for the major crop categories and second for the provinces and special cities of South Korea.

For details of each table, see the notes following Table C-4. Note that zero's indicate entries for which prices or terms-of-trade results were unavailable.

Table C-1.   *Overall Terms of Trade by Crop Category*

(1970 = 1000)

Year	Rice	Barleys	Misc. Grains	Pulses	Pota-toes	Special Crops	Fruits	Vege-tables	Live-stock	All Crops	Year
1959	728	819	823	730	1010	894	1022	920	693	771	1959
1960	810	976	989	832	1074	966	1046	1008	715	856	1960
1961	931	1216	1193	705	1079	780	1023	563	698	894	1961
1962	891	1228	1176	666	1134	1351	1007	565	762	910	1962
1963	1279	1897	1735	993	1630	1330	984	1035	707	1197	1963
1964	1253	1953	1891	1195	1808	1511	1237	773	722	1271	1964
1965	1016	1240	1222	1093	1212	1226	1048	848	971	1048	1965
1966	956	1047	1195	1060	1159	1168	909	922	934	989	1966
1967	929	1094	1236	1236	1185	1069	995	705	1064	967	1967
1968	913	988	1047	775	1051	972	938	574	1111	895	1968
1969	1026	1049	1009	743	1013	917	969	648	954	947	1969
1970	1000	1000	1000	1000	1000	1000	1000	1000	1000	1000	1970
1971	1099	1201	1075	913	1143	1120	1019	817	1027	1048	1971
1972	1223	1367	1222	974	1259	1271	972	726	988	1117	1972
1973	1168	1290	1254	1036	1458	1503	937	671	1067	1106	1973
1974	1255	1218	1242	965	1650	1289	1186	696	969	1123	1974
1975	1286	1374	1281	966	1430	832	1016	772	984	1125	1975

Sources: See notes following Table C-4.

Table C-2. Overall Terms of Trade by Province

(1970 = 1000)

Year	Seoul	Pusan	Kyonggi	Kangwon	North Chung-chong	South Chung-chong	North Jolla	South Jolla	North Kyong-sang	South Kyong-sang	Jeju	South Korea	Year
1959	1141	0	766	838	773	755	748	760	779	759	863	771	1959
1960	1071	0	841	874	849	839	836	837	870	879	979	856	1960
1961	815	0	876	869	899	903	903	883	907	900	880	894	1961
1962	746	817	863	887	926	901	898	911	950	917	985	910	1962
1963	1087	1107	1195	1154	1224	1214	1212	1188	1209	1179	1173	1197	1963
1964	1056	1258	1222	1186	1299	1249	1256	1268	1319	1312	1335	1271	1964
1965	947	1055	1021	1042	1059	1043	1025	1035	1068	1073	1112	1048	1965
1966	918	977	984	1008	1022	1000	978	967	1000	981	1038	989	1966
1967	938	1028	948	1022	939	963	947	968	974	979	1064	967	1967
1968	855	972	873	913	815	893	895	923	888	922	1003	895	1968
1969	872	959	936	930	891	948	958	967	958	943	979	947	1969
1970	1000	1000	1000	1000	1000	1000	1000	1000	1000	1000	1000	1000	1970
1971	971	1008	1003	1016	1017	1041	1063	1080	1056	1059	1071	1048	1971
1972	853	921	1051	1078	1084	1116	1148	1175	1125	1113	1073	1117	1972
1973	938	958	1055	1123	1054	1099	1136	1159	1119	1107	1002	1106	1973
1974	960	973	1086	1104	1066	1132	1169	1166	1137	1138	889	1123	1974
1975	1011	1035	1121	1093	1033	1152	1182	1203	1128	1155	760	1125	1975

Sources: See notes following Table C-4.

Table C-3. Crop Prices and Farm Consumer Deflator

(current won per ton)

1 Rice
2 Barley
3 Naked Barley
4 Wheat
5 Rye
6 Oats
7 Italian Millet
8 Barnyard Millet
9 Glutenous Millet
10 Sorghum
11 Corn
12 Buckwheat
13 Soybean
14 Red Bean

Year	1	2	3	4	5	6	7	8	9	10	11	12	13	14	Year
1959	13767	11534	7163	6457	4670	0	8321	0	0	6525	6026	9140	12120	13108	1959
1960	16423	14741	9569	7778	5674	0	10887	0	0	8341	8271	11169	15019	18304	1960
1961	20369	20226	13216	9229	7885	0	15130	11038	14482	10491	10075	10733	15206	18652	1961
1962	21622	22613	14575	11490	10067	0	16362	9737	16287	11770	10957	14431	16329	17414	1962
1963	34449	39248	25228	17294	14766	0	27452	19650	25785	18384	17064	19587	27232	29319	1963
1964	42868	51548	31660	25359	21516	0	36142	19612	35492	28146	27994	24759	41690	45242	1964
1965	40212	38866	21568	21608	18417	0	27033	12755	23152	21583	19455	24353	43481	49090	1965
1966	42417	36916	19869	23163	18563	0	28278	56243	47662	22977	23638	24578	48892	41430	1966
1967	46726	42674	25359	24627	20483	0	32712	17837	32407	25472	25361	33477	64980	53876	1967
1968	54994	46719	27882	25176	21123	0	33486	18788	34124	26790	27393	31162	45458	60298	1968
1969	68085	54190	34222	24379	21676	0	37425	30945	38065	31434	26404	26262	49560	48429	1969
1970	76490	59091	39006	23830	20527	0	43649	19839	35955	34914	28475	41850	76739	87273	1970
1971	96133	81995	53973	29346	27553	0	57077	15192	27551	42871	33152	54358	79238	91999	1971
1972	122039	104808	70770	38169	33838	0	76234	0	75783	64237	41463	54553	99041	108871	1972
1973	127826	107268	73359	41686	38231	0	84453	0	83955	75060	48719	56763	116278	123748	1973
1974	184762	135566	92130	63803	65435	0	101488	0	100889	107671	70098	64700	143163	144504	1974
1975	233668	190138	128809	82051	80783	0	134317	0	133522	158257	83928	94103	175646	199139	1975

Sources: See notes following Table C-4.

Table C-3. (Continued)

15 Green Bean	22 Cotton
16 Kidney Bean	23 Hemp
17 Peas	24 Ramie
18 Peanut	25 Black Rush
19 Other Bean	26 Paper Mulberry
20 White Potato	27 Wild Sesame
21 Sweet Potato	28 Sesame

Year	15	16	17	18	19	20	21	22	23	24	25	26	27	28	Year
1959	16812	0	0	35408	0	4267	3467	30000	38933	88000	43200	0	17250	50139	1959
1960	31508	0	0	39795	0	4800	4000	37000	54400	104000	25333	0	16525	47417	1960
1961	22561	0	0	38524	15737	5333	4267	38000	50933	106400	22667	0	24530	46499	1961
1962	25688	13786	18273	39795	16730	7200	4800	47000	55733	79200	37867	0	22404	49817	1962
1963	38183	23839	23463	55388	27787	11733	7467	55000	60533	122400	66133	0	34870	85975	1963
1964	50870	34475	40964	79891	42406	15200	11200	73000	82400	148533	86667	0	59126	138256	1964
1965	56330	31606	38358	105282	53594	13600	8267	92000	129600	243467	66667	0	51315	136452	1965
1966	53636	32094	39185	137315	51696	13867	9067	112000	117333	264533	94933	0	66615	177926	1966
1967	75391	38496	35878	140827	65156	16267	10133	120000	134400	251200	116267	0	50171	160869	1967
1968	88716	42630	43206	137233	49130	15467	11467	132000	121067	212533	112267	0	59803	134632	1968
1969	67899	35539	31117	172710	52248	14400	12800	138000	124533	148000	127200	0	72994	276126	1969
1970	135112	51336	53553	187045	82634	16000	14667	220000	113600	191200	355467	0	72961	315586	1970
1971	184623	57261	61596	224066	57238	20800	19200	223000	183467	322667	327733	0	89132	399628	1971
1972	149796	56384	63525	224066	0	28800	23467	283000	209600	40000	320000	0	115482	461766	1972
1973	146777	70565	65141	279550	0	37867	29333	413000	210400	40000	0	0	137225	483993	1973
1974	199090	102259	96646	561382	0	56267	44800	525000	210667	0	0	0	199395	637163	1974
1975	366540	111892	120072	615868	0	52000	50667	508333	0	0	0	0	239178	1007929	1975

Sources: See notes following Table C-4.

Table C-3.   *(Continued)*

29 Caster Bean
30 Pyrethrum Flower
31 Flax
32 Peppermint
33 Hops
34 New Zealand Hemp
35 Other Special Crops

36 Tobacco
37 Rape
38 Loofah
39 Sunflower
40 Tea
41 Mat Rush
42 Sweet Oleander

Year	29	30	31	32	33	34	35	36	37	38	39	40	41	42	Year
1959	12691	0	0	0	0	0	0	57000	0	0	0	0	0	0	1959
1960	11989	0	0	0	0	0	0	62000	23000	0	0	0	0	0	1960
1961	20162	46429	4501	697	3803333	10266	0	75000	23000	0	0	0	0	0	1961
1962	19651	52000	4604	2977	425813	9654	0	133000	23000	0	0	0	0	0	1962
1963	22285	80000	9941	2616	533308	11493	0	156000	40000	63478	54444	19274	54756	0	1963
1964	37499	77879	12342	7593	533250	11733	0	199000	54000	13038	77893	31260	77846	77833	1964
1965	49583	157000	45866	94608	0	16533	108000	157000	52000	107980	49320	48516	66665	131833	1965
1966	51148	150000	53000	52000	813000	18000	99000	161000	62000	50000	52009	52002	95000	99000	1966
1967	54995	135000	48000	54000	736000	16000	385000	164000	50000	600000	55000	53341	116000	100000	1967
1968	60678	132000	47000	52000	718000	16000	94000	167000	58000	583000	53000	52002	113000	98000	1968
1969	56799	139000	25000	1851000	540000	17000	102000	179000	54000	618000	73000	219834	29000	98000	1969
1970	59082	212000	75000	126000	676000	0	147000	236000	55000	145000	66000	65399	51000	0	1970
1971	67989	218000	64000	131000	740000	0	169000	338000	60000	150000	68000	67243	62000	0	1971
1972	74231	0	0	0	0	0	0	437000	79000	0	0	0	79000	0	1972
1973	155439	0	0	0	260000	0	0	472000	100000	1500000	133000	150000	103000	0	1973
1974	178810	0	0	0	260000	0	0	574000	131000	1500000	138000	115000	103000	0	1974
1975	190144	0	0	0	330000	0	0	0	173000	1000000	190000	170000	126000	0	1975

Sources: See notes following Table C-4.

Table C-3. (Continued)

43 Medicinal Crops
44 Pear
45 Persimmon
46 Grapes
47 Peach
48 Orange
49 Other Fruits

50 Plum
51 Radish
52 Carrot
53 Chinese Cabbage
54 Cabbage
55 Welsh Onion
56 Onion

Year	43	44	45	46	47	48	49	50	51	52	53	54	55	56	Year
1959	0	11040	10351	29600	12800	0	0	0	2933	0	5067	7467	5867	7467	1959
1960	0	13013	9784	24800	10933	0	0	0	3467	0	5867	7733	5600	8533	1960
1961	0	13387?	11769	28000	11200	57065	60513	0	2933	0	5067	7467	4533	7467	1961
1962	0	16320	9784	25333	13866	66853	21564	0	3200	14080	3733	8267	8533	14400	1962
1963	0	17547	21056	32267	17600	97065	18784	0	5333	21067	6933	12000	11200	22133	1963
1964	0	29547	24459	46133	22400	97051	28519	0	7467	25867	11200	13067	13600	22667	1964
1965	0	32640	19071	45067	21333	156810	51405	0	5600	30400	7200	17867	19733	25333	1965
1966	0	28960	20206	53067	19467	160020	38040	0	7200	25067	9067	12533	12800	16000	1966
1967	0	36747	26870	61867	29333	204534	34967	0	11467	28800	13867	22400	15467	30400	1967
1968	0	41760	33463	54400	25067	243166	52811	0	10133	42667	12000	17600	22400	31733	1968
1969	0	45707	38213	82400	32800	274380	54741	0	12533	25600	12800	13067	17600	20267	1969
1970	0	60800	53243	78400	31200	291733	61123	0	20800	62933	24533	21600	37600	35733	1970
1971	0	59413	61538	76800	39466	351511	125586	0	17600	56533	20000	33600	35467	40533	1971
1972	0	69600	59270	74133	50315	364000	0	0	14133	49600	16800	22133	22133	20533	1972
1973	0	78827	85643	88533	52157	318133	0	0	18400	63200	22400	28533	32533	34933	1973
1974	0	105600	107976	122933	286933	283733	0	0	29333	82667	28000	57067	46667	43467	1974
1975	0	135947	145056	131467	393600	336800	0	0	40000	84800	44533	46400	60800	65600	1975

Sources: See notes following Table C-4.

Table C-3. (Continued)

57 Garlic
58 Cucumber
59 Squash
60 Korean Melon
61 Water Melon
62 Eggplant
63 Tomato

64 Red Pepper
65 Parsely
66 Spinach
67 Eurdock
68 Taro
69 Other Vegetables
70 Ginger

Year	57	58	59	60	61	62	63	64	65	66	67	68	69	70	Year
1959	34933	0	0	6933	6933	0	5067	49867	0	9333	0	0	0	0	1959
1960	25067	0	0	6667	6400	0	5867	102667	0	7733	0	0	0	0	1960
1961	12267	0	0	5333	5600	0	8533	89067	0	9067	0	0	3809	0	1961
1962	31200	9600	6133	18667	9867	13200	10933	95467	9093	10933	21467	6853	6197	0	1962
1963	105867	13067	7467	20267	16800	13067	21600	225067	11733	17600	26667	9867	17688	0	1963
1964	118400	10933	8533	21067	17333	14133	18133	115200	11467	17067	26133	13867	18930	0	1964
1965	200800	17600	12267	24533	21067	20533	21067	204533	15467	23467	39467	18133	42478	0	1965
1966	143467	18133	10133	21600	19467	17067	18400	382667	18933	22933	36267	27733	31146	133000	1966
1967	101867	18667	10400	21600	20267	16533	22667	210400	20267	28267	41333	28267	28225	62000	1967
1968	121067	16800	10400	23200	20267	12800	23200	218667	36800	36533	48267	25067	30213	50000	1968
1969	126933	17333	11733	28267	22933	14933	29867	305867	23733	30133	45067	29067	30913	98000	1969
1970	97067	18667	13333	37067	36000	17333	26400	786667	49333	50667	67733	42933	48565	99000	1970
1971	145333	29600	20800	39733	38933	27200	22667	646933	46400	52533	63200	41067	64357	134000	1971
1972	199467	33600	26400	51200	47467	46133	40000	668333	40800	40533	72533	47467	0	0	1972
1973	162133	34133	22400	49600	49600	39200	40533	591667	30667	52800	78400	49067	0	172000	1973
1974	134400	53867	31467	61867	58133	58667	60800	938333	53600	73333	101333	67733	0	189867	1974
1975	273867	66933	43733	84533	70667	61867	72000	1220000	61333	86667	103733	67467	0	275200	1975

Sources: See notes following Table C-4.

Table C-3. *(Continued)*

71 Mushroom
72 Asparagus
73 Strawberries
74 Cattle
75 Female Cattle
76 Horse
77 Pig
78 Sheep
79 Goat
80 Rabbit
81 Chicken
82 Duck
83 Cacoon
84 Consumer Deflator

Year	71	72	73	74	75	76	77	78	79	80	81	82	83	84	Year
1959	0	0	0	13296	11872	0	2100	0	877	85	112	0	74400	247	1959
1960	0	0	0	17267	15677	0	2189	0	986	68	114	0	87733	265	1960
1961	0	0	0	18702	17576	6653	2508	0	1154	59	121	66	107467	286	1961
1962	0	0	0	20122	18951	10163	3544	0	1243	125	136	86	137333	317	1962
1963	0	0	0	21772	19977	17433	3362	0	2121	92	154	81	143200	352	1963
1964	0	0	0	25322	23893	15954	4886	0	1053	120	216	158	240000	447	1964
1965	0	0	0	40699	35188	19350	7331	0	1613	184	319	227	304000	517	1965
1966	0	0	0	48094	42318	27083	6529	0	2030	214	355	238	334667	580	1966
1967	0	0	0	61553	54772	31396	8909	0	2820	241	424	278	370400	657	1967
1968	0	0	0	79487	70997	37750	12885	0	3910	262	456	322	0	787	1968
1969	0	0	0	85178	71214	28417	10644	0	4244	258	418	314	417600	867	1969
1970	0	0	0	99496	83676	24416	13578	0	4912	294	505	367	476000	1000	1970
1971	0	0	0	120931	104161	24889	16623	0	5885	335	557	471	570933	1143	1971
1972	0	0	0	159434	142867	24958	15068	0	6502	377	532	0	694133	1304	1972
1973	0	0	0	172028	157238	34167	21338	0	7381	457	618	0	1330133	1430	1973
1974	0	0	0	193211	164627	58333	24632	0	9536	642	888	656	1330133	1924	1974
1975	0	0	0	211983	172580	67500	36294	0	9955	799	1234	925	1330133	2375	1975

Sources: See notes following Table C-4.

Table C-4. Deflated Crop Price Indexes

(1970 = 1000)

1 Rice
2 Barley
3 Naked Barley
4 Wheat
5 Rye
6 Oats
7 Italian Millet

8 Barnyard Millet
9 Glutenous Millet
10 Sorghum
11 Corn
12 Buckwheat
13 Soybean
14 Red Bean

Year	1	2	3	4	5	6	7	8	9	10	11	12	13	14	Year
1959	728	790	743	1096	921	0	771	0	0	756	856	884	639	608	1959
1960	810	941	925	1231	1043	0	941	0	0	901	1096	1007	738	791	1960
1961	931	1196	1184	1354	1343	0	1211	1945	1408	1050	1237	896	692	747	1961
1962	891	1207	1178	1521	1547	0	1182	1548	1428	1063	1213	1087	671	629	1962
1963	1279	1886	1837	2061	2043	0	1786	2813	2037	1495	1702	1329	1008	954	1963
1964	1253	1951	1815	2380	2344	0	1852	2211	2208	1803	2199	1323	1215	1159	1964
1965	1016	1272	1069	1753	1735	0	1197	1243	1245	1195	1321	1125	1095	1087	1965
1966	956	1077	878	1675	1559	0	1116	4887	2285	1134	1431	1012	1098	818	1966
1967	929	1099	989	1572	1518	0	1140	1368	1371	1110	1355	1217	1288	939	1967
1968	913	1004	908	1342	1307	0	974	1203	1205	974	1222	946	752	877	1968
1969	1026	1057	1011	1179	1217	0	988	1799	1221	1038	1069	723	744	640	1969
1970	1000	1000	1000	1000	1000	0	1000	1000	1000	1000	1000	1000	1000	1000	1970
1971	1099	1213	1210	1077	1174	0	1144	669	670	1074	1018	1136	903	922	1971
1972	1223	1360	1391	1228	1264	0	1339	0	1616	1410	1116	999	989	956	1972
1973	1168	1269	1315	1223	1302	0	1353	0	1632	1503	1196	948	1059	991	1973
1974	1255	1192	1227	1391	1656	0	1208	0	1458	1602	1279	803	969	860	1974
1975	1286	1354	1390	1449	1656	0	1295	0	1563	1908	1241	946	963	960	1975

Sources: See notes following this table.

Table C-4. (Continued)

15 Green Bean
16 Kidney Bean
17 Peas
18 Peanut
19 Other Bean
20 White Potato
21 Sweet Potato

22 Cotton
23 Hemp
24 Ramie
25 Black Rush
26 Paper Mulberry
27 Wild Sesame
28 Sesame

Year	15	16	17	18	19	20	21	22	23	24	25	26	27	28	Year
1959	503	0	0	766	0	1079	956	552	1387	1863	492	0	957	643	1959
1960	879	0	0	802	0	1132	1029	634	1807	2052	268	0	854	566	1960
1961	583	0	0	720	665	1165	1017	603	1567	1945	222	0	1175	515	1961
1962	599	847	1076	671	638	1419	1032	673	1513	1306	336	0	968	497	1962
1963	802	1319	1244	841	955	2083	1446	710	1622	1818	528	0	1357	773	1963
1964	842	1502	1711	955	1148	2125	1708	742	2206	1737	545	0	1812	980	1964
1965	806	1190	1385	1088	1254	1644	1090	808	1780	2462	362	0	1360	836	1965
1966	684	1077	1261	1265	1078	1494	1065	877	1800	2385	460	0	1574	972	1966
1967	849	1141	1019	1145	1200	1547	1051	830	1354	1999	497	0	1046	775	1967
1968	834	1055	1025	932	755	1228	993	762	1264	1412	401	0	1041	542	1968
1969	579	798	670	1065	729	1038	1006	723		892	412	0	1153	1009	1969
1970	1000	1000	1000	1000	1000	1000	1000	1000	1000	1000	1000	0	1000	1000	1970
1971	1195	975	1006	1048	606	1137	1145	886	1412	1476	806	0	1068	1107	1971
1972	850	842	909	918	0	1380	1226	986	1414	1604	690	0	1213	1122	1972
1973	759	961	850	1045	0	1655	1398	1312	1295	1462	0	0	1315	1072	1973
1974	765	1035	937	1559	0	1827	1587	1240	963	0	0	0	1420	1049	1974
1975	1142	917	944	1386	0	1368	1454	972	0	0	0	0	1380	1344	1975

Table C-4.   *(Continued)*

29 Caster Bean
30 Pyrethrum Flower
31 Flax
32 Peppermint
33 Hops
34 New Zealand Hemp
35 Other Special Crops

36 Tobacco
37 Rape
38 Loofah
39 Sunflower
40 Tea
41 Mat Rush
42 Sweet Oleander

Year	29	30	31	32	33	34	35	36	37	38	39	40	41	42	Year
1959	869	0	0	0	0	0	0	977	0	0	0	0	0	0	1959
1960	765	0	0	0	0	0	0	991	1578	0	0	0	0	0	1960
1961	1193	765	209	19	19672	0	0	1111	1462	0	0	0	0	0	1961
1962	1049	773	193	74	1987	0	0	1777	1319	0	0	0	0	0	1962
1963	1071	1072	376	58	2241	0	0	1877	2066	1243	2343	837	3050	0	1963
1964	1419	821	368	134	1764	0	0	1886	2196	201	2640	1069	3414	0	1964
1965	1623	1432	1182	1452	0	0	1421	1286	1828	1440	1445	1434	2528	0	1965
1966	1492	1219	1218	711	2073	0	1161	1176	1943	594	1358	1370	3211	0	1966
1967	1416	969	974	652	1657	0	3986	1057	1383	6298	1268	1241	3461	0	1967
1968	1304	791	796	524	1349	0	812	899	1339	5108	1020	1010	2815	0	1968
1969	1108	756	384	16944	921	0	800	874	1132	4915	1275	3877	655	0	1969
1970	1000	1000	1000	1000	1000	0	1000	1000	1000	1000	1000	1000	1000	0	1970
1971	1006	899	746	909	957	0	1005	1253	954	905	901	899	1063	0	1971
1972	963	0	0	0	0	0	0	1420	1101	0	0	0	1187	0	1972
1973	1839	0	0	0	268	0	0	1398	1271	7234	1409	1603	1412	0	1973
1974	1573	0	0	0	199	0	0	1264	1237	5376	1086	913	1049	0	1974
1975	1355	0	0	0	205	0	0	0	1324	2903	1212	1094	1040	0	1975

Sources: See notes following this table.

Table C-4. (Continued)

43 Medicinal Crops
44 Pear
45 Persimmon
46 Grapes
47 Peach
48 Orange
49 Other Fruits

50 Plum
51 Radish
52 Carrot
53 Chinese Cabbage
54 Cabbage
55 Welsh Onion
56 Onion

Year	43	44	45	46	47	48	49	50	51	52	53	54	55	56	Year
1959	0	735	787	1528	1660	0	0	0	570	0	836	1399	631	845	1959
1960	0	807	693	1193	1322	0	0	0	628	0	902	1350	562	901	1960
1961	0	769	772	1248	1255	683	3461	0	493	0	722	1208	421	730	1961
1962	0	846	579	1019	1401	722	1112	0	485	705	480	1207	715	1271	1962
1963	0	819	1123	1169	1602	945	873	0	728	950	802	1578	846	1759	1963
1964	0	1087	1027	1316	1606	744	1043	0	803	919	1021	1353	809	1419	1964
1965	0	1038	692	1111	1322	1039	1626	0	520	934	567	1599	1015	1371	1965
1966	0	821	654	1167	1075	945	1073	0	596	686	637	1000	586	772	1966
1967	0	919	768	1201	1430	1067	870	0	839	696	860	1578	626	1294	1967
1968	0	872	798	881	1020	1059	1097	0	618	861	621	1035	756	1128	1968
1969	0	867	827	1212	1212	1084	1032	0	694	469	601	697	539	654	1969
1970	0	1000	1000	1000	1000	1000	1000	0	1000	1000	1000	1000	1000	1000	1970
1971	0	854	1011	857	1106	1054	1797	0	740	785	713	1360	825	992	1971
1972	0	877	853	725	1236	956	0	0	521	604	525	785	451	440	1972
1973	0	906	1124	789	1169	762	0	0	618	702	638	923	605	683	1973
1974	0	902	1054	814	4779	505	0	0	732	682	593	1373	645	632	1974
1975	0	941	1147	706	5311	486	0	0	809	567	764	904	680	772	1975

Sources: See notes following this table.

Table C-4. (Continued)

57 Garlic
58 Cucumber
59 Squash
60 Korean Melon
61 Water Melon
62 Eggplant
63 Tomato

64 Red Pepper
65 Parsely
66 Spinach
67 Burdock
68 Taro
69 Other Vegetables
70 Ginger

Year	57	58	59	60	61	62	63	64	65	66	67	68	69	70	Year
1959	1457	0	0	757	779	0	777	256	0	745	0	0	0	0	1959
1960	974	0	0	678	670	0	838	492	0	575	0	0	0	0	1960
1961	441	0	0	503	543	0	1130	395	0	625	0	0	274	0	1961
1962	1013	1622	1451	1588	864	2402	1306	382	581	680	999	503	402	0	1962
1963	3098	1988	1591	1553	1325	2141	2324	812	675	986	1118	652	1034	0	1963
1964	2728	1310	1431	1271	1077	1824	1536	327	519	753	863	722	871	0	1964
1965	4001	1823	1779	1280	1131	2291	1543	502	606	895	1127	816	1691	0	1965
1966	2548	1674	1310	1004	932	1697	1201	838	661	780	923	1113	1105	2316	1966
1967	1597	1522	1187	886	856	1451	1306	407	625	849	928	1002	884	953	1967
1968	1584	1143	991	795	715	938	1116	353	947	916	905	741	790	641	1968
1969	1508	1070	1014	879	734	993	1304	448	554	685	767	780	734	1141	1969
1970	1000	1000	1000	1000	1000	1000	1000	1000	1000	1000	1000	1000	1000	1000	1970
1971	1309	1387	1364	937	946	1372	751	719	822	907	816	836	1159	1184	1971
1972	1575	1380	1518	1059	1011	2041	1161	651	634	613	821	847	0	0	1972
1973	1168	1278	1174	935	963	1581	1073	525	434	728	809	799	0	1214	1973
1974	719	1499	1226	867	839	1759	1196	619	564	752	777	819	0	996	1974
1975	1187	1509	1381	960	826	1502	1148	652	523	720	644	661	0	1170	1975

Sources: See notes following this table.

Table C-4. (Continued)

71 Mushroom
72 Asparagus
73 Strawberries
74 Cattle
75 Female Cattle
76 Horse
77 Pig

78 Sheep
79 Goat
80 Rabbit
81 Chicken
82 Duck
83 Cacoon
84 Consumer Deflator

Year	71	72	73	74	75	76	77	78	79	80	81	82	83	84	Year
1959	0	0	0	541	574	0	626	0	722	1170	897	0	632	247	1959
1960	0	0	0	654	706	0	608	0	757	870	851	0	695	265	1960
1961	0	0	0	657	734	952	645	0	821	700	837	626	789	286	1961
1962	0	0	0	637	714	1313	823	0	798	1340	849	738	910	317	1962
1963	0	0	0	621	678	2028	703	0	1226	887	865	626	854	352	1963
1964	0	0	0	569	638	1461	804	0	479	911	956	961	1127	447	1964
1965	0	0	0	791	813	1532	1044	0	634	1207	1221	1196	1235	517	1965
1966	0	0	0	833	871	1912	828	0	712	1251	1211	1117	1212	580	1966
1967	0	0	0	941	996	1957	998	0	873	1244	1277	1152	1184	657	1967
1968	0	0	0	1015	1078	1964	1205	0	1011	1129	1146	1114	0	787	1968
1969	0	0	0	987	981	1342	904	0	996	1010	954	986	1011	867	1969
1970	0	0	0	1000	1000	1000	1000	0	1000	1000	1000	1000	1000	1000	1970
1971	0	0	0	1063	1089	891	1071	0	1048	996	964	1122	1049	1143	1971
1972	0	0	0	1228	1309	783	851	0	1015	982	805	0	1118	1304	1972
1973	0	0	0	1209	1314	978	1098	0	1050	1085	855	0	1954	1430	1973
1974	0	0	0	1009	1022	1241	942	0	1008	1132	912	926	1452	1924	1974
1975	0	0	0	897	868	1164	1125	0	853	1142	1027	1059	1176	2375	1975

Sources: See notes following this table.

Notes to Tables in Appendix C

Table C-1

The data in this table were obtained by dividing total South Korean output for each category and year valued in current prices by the corresponding output valued in constant 1970 prices. Each figure, then, represents the degree to which the cultivation of crops in that category were more or less remunerative in any year when compared to the remuneration from the same output sold at 1970 prices. The prices used to make the calculations are given in Table C-3.

Table C-2

The data in this table are similar to those in Table C-1, except that the comparison is for the combined output of all crops for any single province or special city. This table, then, is nothing more than the data in Table B-17 divided by corresponding data in Table B-16. The results, as are those in Table C-1, are a Paasche index of prices because they compare output measured in prices from the same year to the same output valued in base-year prices. The prices used to make the value calculations are given in Table C-3.

Table C-3

These prices were converted to won per metric ton from price data in two editions of Nong-eop Hyeop-dong Cho-hap Chung-ang Hwei Chosa-bu [National Agricultural Cooperative Association, Research Division], *Nong-chon Mul-ka Chong-ram* [Digest of Farm Village Prices], Seoul: N.A.C.A., 1971 and 1976 (in Korean). The conversion from Korean volume and weight units to metric tons was done using information in Republic of Korea, Ministry of Agriculture and Forestry, *Yearbook of Agriculture and Forestry Statistics,* 1975, pp. 459-461. A zero indicates price data were unavailable for that entry.

Table C-4

This table compares the change of individual crop prices with the change in the farm consumer price index. Prices and the consumer index were taken from Table C-3, and each entry in this table represents the result of dividing the corresponding price by the consumer deflator for that year.

BIBLIOGRAPHY AND INDEX

BIBLIOGRAPHY

Ban, Sung-hwan, *Hangug Nongeop ui Seongchang* [Growth of Korean Agriculture], Seoul: KDI Press, 1974. (in Korean)

Bank of Chosun, *Economic History of Chosun,* Seoul: Bank of Chosun, 1921.

Bank of Chosun, "Chosun Ilbonin Tochi Soyu ui Teoksu Gwa Ke Cheobun Munje" [Characteristics of Former Japanese Land Ownership], in Bank of Chosun, *Monthly Statistical Review,* No. 15, August and September 1948, pp. 125-135. (in Korean)

Bank of Chosun, *Chosun Kyeongje Yeonbo* [Chosun Economic Yearbook], Seoul: Bank of Chosun, 1948. (in Korean)

Bank of Korea, *Economic Statistics Yearbook,* Seoul: Bank of Korea, annual.

Bank of Korea, *Saneop Chonggam* [Industrial Almanac], Seoul, 1954. (in Korean)

Choi, Ho-chin, *The Economic History of Korea,* Seoul: The Freedom Library, 1971.

Chosen Sotoku-fu [Chosun Government General], *Chosen no Sosaku Kanshu* [The Practice of Tenancy in Chosun], Seoul: Chosen Sotoku-fu, 1928(?). (in Japanese)

Chosen Sotoku-fu [Chosun Government General], *Nogyo Tokei Hyo* [Agricultural Statistical Tables], Seoul, 1934. (in Japanese)

Chosen Sotoku-fu [Chosun Government General], *Showa 8-Nen Nogyo Tokei Hyo* [1933 Agricultural Statistical Tables], Seoul, 1937. (in Japanese)

Chosen Sotoku-fu [Chosun Government General], *Tochi Kairyo Jigyo no Gairyo* [The General State of Land Improvement Works], Seoul, 1932. (in Japanese)

Chosen Sotoku-fu [Chosun Government General], *Tokei Nenbo* [Statistical Yearbook], Seoul: Chosun Government General, annual 1910-1942. (in Japanese)

Chosen Sotoku-fu, Tetsu-do Kyoku [Chosun Government General, Railroad Department], *Chosen Tetsu-do Rosen* [Collected Materials on the Chosun Railroads], Seoul, 1929(?). (in Japanese)

Chung, Young-il, "Over-time Changes in the Regional and Urban-rural Income Differences in Korea," Kunitachi, Japan: Institute of Economic Research, Hitotsubashi University, May 1976. (mimeographed)

_____ , "Kyeong-chi Myeon-cheok ui Chu-kye wa Bun-seok (1911-1971)" [Estimation and Analysis of Arable Land Area (1911-1971)], *The Korean Economic Journal,* June 1975. (in Korean)

Darwent, D. F., "Growth Poles and Growth Centers in Regional Planning: A Review," *Environment and Planning,* Vol. 1, 1969, pp. 5-31. Reprinted in W. Alonso and J. Friedman (eds.), *Regional Policy: Readings in Theory and Applications,* Cambridge: M.I.T. Press, 1975.

Eitsuke, Zenski, "Chosun ui Ingu Tongei" [Chosun Population Statistics] in Bureau of Statistics, Economic Planning Board. *Hangug ui Ingu Dongtae Tongei* [Vital Statistics of Korea], Seoul, 1965. (in Korean)

Hishimoto, Choji, *Chosen Mai no Kenkyu* [A Study of Chosun Rice], Tokyo: Chikura Shobo, 1938. (in Japanese)

Ho, Yhi-min, "Korean Rice, Taiwan Rice, and Japanese Agricultural Stagnation: An Economic Consequence of Colonialism," Paper No. 16, Program of Development Studies, Rice University, Summer 1971.

Isard, Walter, *General Theory; Social, Political, Economic and Regional,* Cambridge: M.I.T. Press, 1969.

Ishikawa, Shigeru, *Chosen Nogyo Seisankaku no Shukei, Senzen no Bu* [Collected Statistics of Chosun Agricultural Production, Pre-war Part], Kunitachi, Japan: Economic Research Center, Hitotsubashi University, 1973. (mimeographed, in Japanese)

Juhn, Daniel Sungil, "Entrepreneurship in an Undeveloped Economy: The Case of Korea, 1980-1940," Ph. D Dissertation, D.B.A., George Washington University, 1965.

Kim, Gwang-seop (ed.), *Tong-nip Shin-moon, Ch'ook-Swae-P'an* [The Independent Newspaper, Reduced Size Republication], Seoul: original English and Korean texts, Vol. 1, No. 1, April 7, 1896 to

Vol. 1, No. 116, December 31, 1896)

Kim, Jo-tai, *Chosun Michak Yeongu* [Research in Chosun Rice], Seoul: Jeong-in Sa, 1948. (in Korean)

Kim, Sa-hon, "Kwanggong-eop Senseosu Charyo" [Statistics from Industrial Censuses], Seoul, 1975. (in Korean, unpublished)

Komikawa, Kuro (ed.), *Chosen Nogyo Hattatsu-shi* [The History of the Chosun Agriculture], Two volumes: *Seisaku-bu* [Policy Volume] and *Hattatsu-bu* [Development Volume], Tokyo: Chosen Nogyo-sha. 1944 (in Japanese)

Kwon, Tai-hwan, *The Population of Korea,* Seoul: Seoul National University, 1975.

Mizoguchi, Toshiyuki, *Taiwan Chosen no Keizai Seicho* [The Economic Growth of Taiwan and Chosun], Tokyo: Iwanami Shoten, 1975. (in Japanese)

Moon, Pal-yong, *Nongsanmul Kakyeok Bunseok Non* [Analysis of Farm Product Prices], Seoul: KDI Press, 1975. (in Korean)

Nicholls, William H., "Industrialization, Factor Markets, and Agricultural Development," *Journal of Political Economy,* Vol. 69, No. 4, August 1961, pp. 319-340.

_____ , "The Transformation of Agriculture in a Semi-Industrialized Country: The Case of Brazil," in Erik Thorbecke (ed.), *The Role of Agriculture in Economic Development,* New York: National Bureau of Economic Research, 1969.

Nong-eop Hyeop-dong Cho-hap Chung-ang Hwei Chosa-bu [National Agricultural Cooperative Association, Research Division], *Nong-chon Mul-ka Chong-ram* (1959-1970) [Digest of Farm Village Prices], Seoul: N.A.C.A., 1971. (in Korean)

Nong-eop Hyeop-dong Cho-hap Chung-ang Hwei Chosa-bu [National Agricultural Cooperative Association, Research Division], *Nong-chon Mul-ka Chong-ram (1959. 1-1974. 6)* [A Digest of Farm Village Prices], Seoul, 1974. (mimeographed, in Korean)

Ogawa, Ichishi and Seiichi Higashihata, *Chosen Maikoku Keizairon* [The Economics of Chosun Rice], Tokyo: Japanese Technical Society, 1935. (in Japanese)

Osgood, Cornelius, *The Koreans and Their Culture,* Rutland: Charles E. Tuttle, 1951.

Republic of Korea, Central Meteorological Office, *Climatic Tables of Korea (1931-1960), Climatological Standard Normals, Part 1,* Seoul, 1968. (in Korean)

Republic of Korea, Economic Planning Board, *Korea Statistical Year-*

book, Seoul: Economic Planning Board, annual.

Republic of Korea, Ministry of Agriculture, and Fisheries, *Agricultural Census, 1970,* Seoul, 1974.

Republic of Korea, Ministry of Agriculture and Fisheries, (Ministry of Agriculture and Forestry prior to March 1, 1973) *Statistical Yearbook,* Seoul, annual.

Republic of Korea, Ministry of Agriculture and Fisheries, *Farm Household Economy Survey Report, 1975,* Seoul, 1976.

Republic of Korea, Ministry of Home Affairs, "Value added by County in Farming, Fishing and Forestry Households, 1967-1970." Seoul, 1972. (mimeographed, in Korean)

Ruttan, Vernon W., "The Impact of Urban Industrial Development on Agriculture in the Tennessee Valley and the Southeast," *Journal of Farm Economics,* Vol. 37, No. 1, February 1955, pp. 38-56. Reprinted in American Economic Association, *Readings in the Economics of Agriculture,* Homewood: Richard D. Irwin, 1969, pp. 339-358.

Sakurai, Hiroshi, *Kankoku Tochi Kaikaku no Saikento* [A Reexamination of South Korean Land Reform], Tokyo: Asian Economic Research Center, 1976. (in Japanese)

Schultz, Theodore W., *The Economic Organization of Agriculture,* New York: McGraw-Hill, 1953.

_____ , "A Framework for Land Economics—The Long View," *Journal of Farm Economics,* Vol. 33, No. 2, May 1951, pp. 204-215.

_____ , "Reflections on Poverty within Agriculture," *Journal of Political Economy,* Vol. 58, No. 1, February 1950, pp. 1-15. Reprinted in American Economic Association, *Readings in the Economics of Agriculture,* Homewood: Richard D. Irwin, 1969, pp. 321-338.

Shaw, Lawrence H., "The Effects of Weather on Agricultural Output: A Look at Methodology," *Journal of Farm Economics,* Vol. 46, No. 1, February 1964, pp. 218-230.

Suh, Sang-chul, "*Growth and Structural Changes in the Korean Economy Since 1910,*" Ph. D Dissertation, Harvard University, 1965.

Suzuki, Takeo, *Chosen no Keizai* [Economy of Chosun], Tokyo: Nihon Hyoron-sha, 1942. (in Japanese)

Tang, Anthony M., *Economic Development in the Southern Piedmont, 1860-1950: Its Impact on Agriculture,* Chapel Hill: University of North Carolina Press, 1958.

Theil, Henri, *Economics and Information Theory,* Chicago: Rand McNally, 1967.

Yamaguchi, Sei (ed.), *Chosen Sangyoshi* [Chosun Industrial Report], Tokyo: Oha Kuro, 1910. (in Japanese)

INDEX

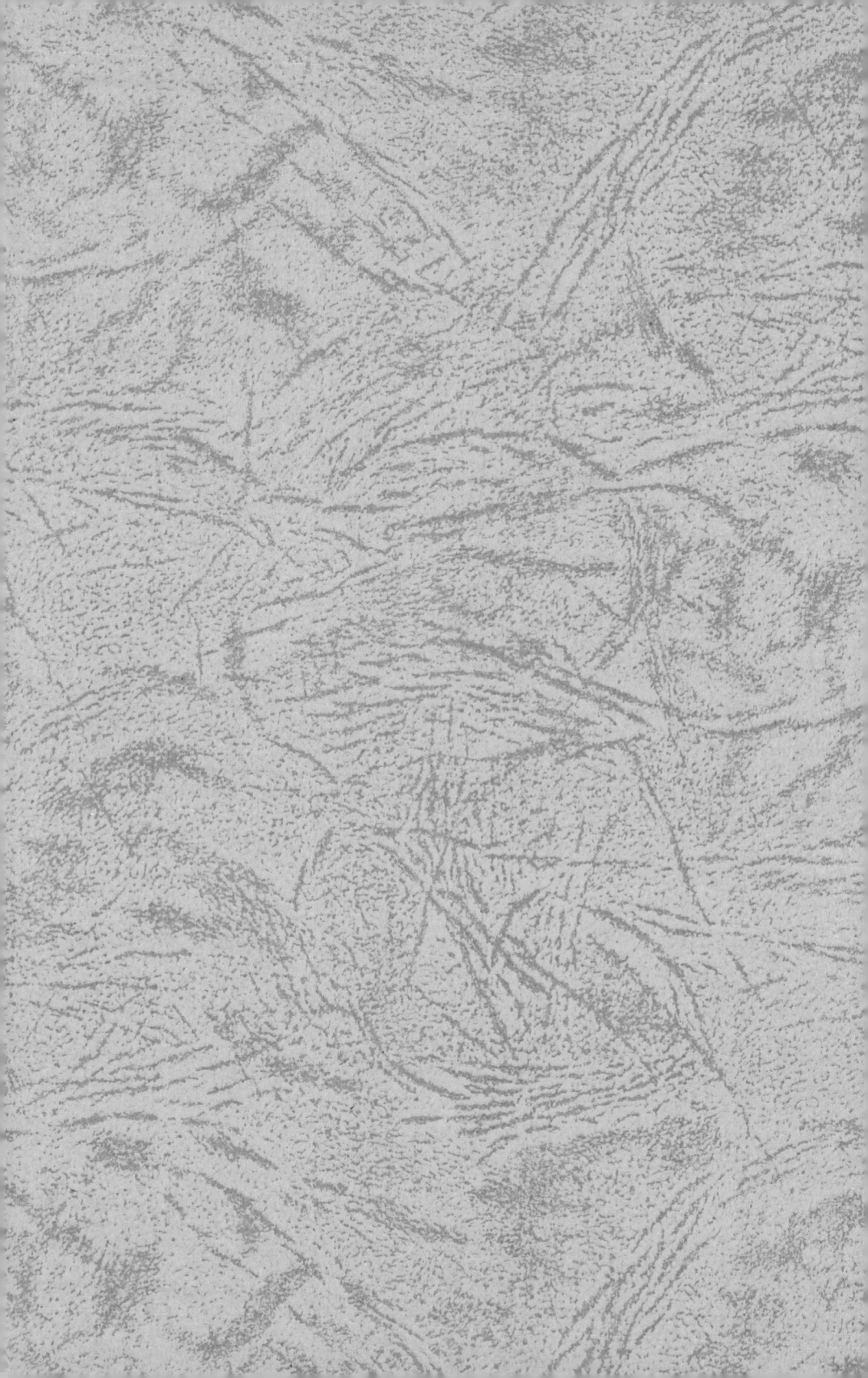